JAMESTOWN'S

Number Power

Jerry Howett

JAMESTOWN PUBLISHERS

a division of NTC/CONTEMPORARY PUBLISHING GROUP
Lincolnwood, Illinois USA

ISBN: 0-8092-2281-7

Published by Jamestown Publishers,
a division of NTC/Contemporary Publishing Group, Inc.,
4255 West Touhy Avenue,
Lincolnwood (Chicago), Illinois 60712-1975 U.S.A.
© 2001 by NTC/Contemporary Publishing Group, Inc.

1 2 3 4 5 6 7 8 9 VH 16 15 14 13 12 11 10 9 8 7 6 5 4 3 2 1

Table of Contents

To the Student   vi

Pretest   1

BUILDING NUMBER POWER   5

Place Value
Understanding Place Value   6
Zeros and Place Value   9
Rounding Whole Numbers   11

Addition
Addition Skills Inventory   13
Basic Addition Facts   15
Adding Larger Numbers   17
Adding and Carrying   19
Adding Numbers Written Horizontally   23
Addition Shortcuts   24
Rounding, Estimating, and Using a Calculator   25
Applying Your Addition Skills   27
Addition Review   31

Subtraction
Subtraction Skills Inventory   33
Basic Subtraction Facts   35
Subtracting Larger Numbers   37
Subtracting and Borrowing   39
Subtracting Numbers Written Horizontally   44
Subtracting from Zeros   45
Subtraction Shortcuts   49
Rounding, Estimating, and Using a Calculator   50

OS-OK

Applying Your Subtraction Skills 52

Subtraction Review 57

Multiplication

Multiplication Skills Inventory 60

Basic Multiplication Facts 62

The Multiplication Table 65

Multiplying by One-Digit Numbers 66

Multiplying by Larger Numbers 67

Multiplying Numbers Written Horizontally 69

Multiplying and Carrying 71

Multiplying by 10, 100, and 1,000 74

Rounding, Estimating, and Using a Calculator 76

Applying Your Multiplication Skills 79

Multiplication Review 84

Division

Division Skills Inventory 87

Basic Division Facts 90

Dividing by One-Digit Numbers 93

Dividing with Remainders 96

Mental Math and Properties of Numbers 99

Dividing by Two-Digit Numbers 100

Dividing by Three-Digit Numbers 105

Rounding and Estimating 107

Two-Digit Accuracy and Using a Calculator 109

Applying Your Division Skills 111

Division Review 117

Posttest A 120

Posttest B 125

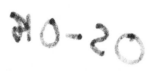

USING NUMBER POWER 129

Changing Units of Measurement 130

Adding Measurements 131

Subtracting Measurements 132

Multiplying Measurements 133

Dividing Measurements 134

Perimeter: Measuring the Distance Around a Rectangle 135

Area: Measuring the Space Inside a Rectangle 137

Volume: Measuring the Space Inside a
3-Dimensional Object 140

Pricing a Meal from a Menu 142

Reading Sales Tax Tables 143

Checking Your Change with Sales Receipts 145

Using a Calorie Chart 147

Finding an Average 149

Reading Paycheck Stubs 151

Using a Table to Look Up Check Cashing Rates 153

Understanding Tax Statements 155

Figuring the Cost of Electricity 157

Reading Time Schedules 159

Renting a Car 161

Using a Calculator 162

Using Mental Math 163

Using Estimation 165

Units of Measurement 168

Glossary 169

Index 172

To The Student

This Number Power book is designed to help you build and use the basic whole number math skills found in the workplace and in everyday experiences.

The first section of the book, called Building Number Power, provides step-by-step instruction and plenty of practice in addition, subtraction, multiplication, and division of whole numbers. Work in these chapters begins with a skills inventory to pinpoint your strengths and weaknesses. Each chapter ends with a cumulative review to measure your progress and to show you which areas need additional work.

Scattered through the book are tips on using a calculator, estimating answers, and solving problems with mental math shortcuts. The following icons will alert you to problems where using these skills will be especially helpful.

 calculator icon

 estimation icon

 mental math icon

The section called Using Number Power will give you a chance to put your skills to work by applying them in real-life problems.

Inside the back cover is a chart to help you keep track of your score on each exercise. Also included in the back are quick reference pages for using a calculator, estimation, and mental math. Refer to these pages for a quick review and helpful explanation.

The whole number skills that you learn, practice, and apply in this book are the building blocks for all of mathematics. Work your way slowly and carefully through this book to build a solid foundation for your increasing Number Power.

Pretest

This test will tell you which sections of *Addition, Subtraction, Multiplication, and Division* you need to concentrate on. Do every problem that you can. After you check your answers, the chart at the end of the test will guide you to the pages of the book where you need work.

1. Which digit in the number 27,346 is in the *hundreds* place?

2. The area of Greece is 50,949 square miles. Round the area to the nearest *thousand* square miles.

3. Which digit in the number 2,476,500 is in the *ten thousands* place?

4. The town of Hudson Village has 3,468 registered voters. What is the number of registered voters rounded to the nearest hundred?

5. 46
   + 52

6.   908
    3,470
   +   65

7. $3 + 26 + 47 + 4 =$

8. $21,859 + 9,786 =$

9. $258,497 + 87,063 =$

10. Round 923 to the nearest *hundred*. Round 8,609 to the nearest *thousand*. Round 38,519 to the nearest *ten thousand*. Then add the rounded numbers.

11. The driving distance from Los Angeles to San Francisco is 403 miles. The distance from San Francisco to Portland, Oregon, is 652 miles. The distance from Portland to Seattle is 175 miles. Find the total distance from Los Angeles to Seattle by way of San Francisco and Portland.

12. A concerned citizens group received a $15,000 grant from the federal government to improve a playground. They also received a $12,500 grant from the state and $9,487 in private donations. Find the total amount of funds the group has raised for playground improvements.

13.   167
    − 52

14.   6,324
    − 4,295

15. $1,827 - 959 =$

16. $20,000 - 6,123 =$

17. $803,400 - 29,753 =$

18. Round each number to the nearest *hundred*. Then subtract the rounded numbers.

    $2,981 - 746$

19. In a recent year, the average price for an existing single-family house in El Paso was $76,300. The average price for an existing single-family house in Dallas was $101,200. How much more was the average price of an existing house in Dallas than the price of an existing house in El Paso?

20. In a 10-year period, the number of farm workers in the U.S. dropped from 4,132,000 to 2,864,000. By how many workers did the number drop?

21.   54
        $\times\,21$

22.   358
        $\times\;\;\,9$

23. $(76)(54) =$

24. $100 \cdot 258 =$

25. $37(1,925) =$

26. Round 82 to the nearest *ten*. Round 9,271 to the nearest *thousand*. Then find the product of the rounded numbers.

27. Steve's sport-utility vehicle gets 13 miles on 1 gallon of gasoline in the city. The gasoline tank holds 30 gallons. How many miles of city driving can Steve get on a full tank of gasoline?

28. Sound travels at a speed of 1,128 feet per second. Find, to the nearest *hundred* feet, the distance a thunderbolt travels in 20 seconds.

29. $8\overline{)536}$

30. $7\overline{)5,645}$

31. $1,288 \div 23 =$

32. $\dfrac{51,100}{70} =$

33. $8,313/163 =$

34. Find the answer to the problem $69,736 \div 92$ to the nearest *ten*.

35. Diana works for a newspaper delivery service. Sunday papers are packaged together with 15 papers in each bundle. How many bundles are required to package the 3,000 papers that are delivered each Sunday to the suburb of Grey Gardens?

36. One year a leading player for the National Basketball Association scored 1,886 points during the season. He played a total of 82 games. Find the average number of points he scored per game.

Pretest Chart

If you miss more than one problem in any section of this test, you should complete the lessons on the practice pages indicated on this chart. If you miss only one problem in a section of this test, you may not need further study in that chapter. However, before you skip those lessons, we recommend that you complete the review test at the end of that chapter. For example, if you miss one problem about addition, you should pass the Addition Review (pages 31–32) before beginning the chapter on subtraction. This longer inventory will be a more precise indicator of your skill level.

PROBLEM NUMBERS	SKILL AREA	PRACTICE PAGES
1, 2, 3, 4	place value	6–12
5, 6, 7, 8, 9, 10, 11, 12	addition	15–30
13, 14, 15, 16, 17, 18, 19, 20	subtraction	35–56
21, 22, 23, 24, 25, 26, 27, 28	multiplication	62–83
29, 30, 31, 32 33, 34, 35, 36	division	90–116

Building
Number
Power

PLACE VALUE

Understanding Place Value

What do these four numbers have in common?

    4,321   1,234   3,412   2,143

You probably noticed that they are all four-**digit** numbers, but did you notice that all four numbers are made up of the same digits: 1, 2, 3, and 4? The digits are the same, but each number has a different value. This is because the digits are in different **places** in each number. In our number system the place of the digit tells you its value. In other words, each digit in a number has a **place value.**

In the box at the right, the four numbers are written with each digit under the name of the place in which it stands.

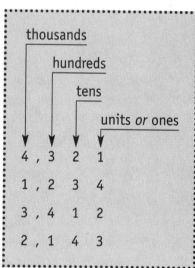

EXAMPLE   What is the value of the digit 4 in each of the numbers at the top of this page?

    The digit 4 in 4,321 is in the thousands place. It has a value of 4,000.

    The digit 4 in 1,234 is in the units or ones place. It has a value of 4.

    The digit 4 in 3,412 is in the hundreds place. It has a value of 400.

    The digit 4 in 2,143 is in the tens place. It has a value of 40.

Use the number 7,856 to answer questions 1 to 6.

1. Which digit is in the thousands place?

2. Which digit is in the hundreds place?

3. What is the value of the digit 7?

4. What is the value of the digit 8?

5. What is the value of the digit 5?

6. What is the value of the digit 6?

Use the number 9,204 to answer questions 7 to 12.

7. Which digit is in the hundreds place?

8. Which digit is in the units place?

9. What is the value of the digit 9?

10. What is the value of the digit 2?

11. What is the value of the digit 0?

12. What is the value of the digit 4?

There are 5,280 feet in 1 mile. Use the number 5,280 to answer questions 13 to 16.

13. What is the value of the digit 5?

14. Which digit is in the hundreds place?

15. Which digit is in the tens place?

16. Which digit is in the units place?

Serena paid $6,750 for a used car. Use the number 6,750 to answer questions 17 to 19.

17. What is the value of the digit 7?

18. Which digit is in the thousands place?

19. Which digit is in the units place?

Mr. Johnson was born in 1924. Use the number 1924 to answer questions 20 to 22.

20. What is the value of the digit 2?

21. Which digit is in the thousands place?

22. Which digit is in the hundreds place?

The cost for each pupil at Greendale School is $3,614. Use the number 3,614 to answer questions 23 to 25.

23. What is the value of the digit 6?

24. Which digit is in the thousands place?

25. Which digit is in the units place?

In his professional baseball career, Hank Aaron played in 3,298 games. Use the number 3,298 to answer questions 26 to 29.

26. What is the value of the digit 2?

27. Which digit is in the thousands place?

28. Which digit is in the tens place?

29. Which digit is in the units place?

30. Which of the following numbers has a 5 in the hundreds place?

    a. 6,235
    b. 5,623
    c. 3,265
    d. 2,536

Zeros and Place Value

What do these three numbers have in common?

    6,200   6,020   6,002

You probably noticed that each number has four digits and also that each number is written with the digits 0, 2, and 6. The numbers are different because zeros hold different places in each number.

In 6,200 zeros hold the tens and units places.

1. In 6,020 zeros hold the _____ and _____ places.

2. In 6,002 zeros hold the _____ and _____ places.

3. In 9,000 zeros hold the _____, _____, and _____ places.

4. In $400 zeros hold the _____ and _____ places.

The chart below lists the first seven places in the whole number system. Below the place names are three examples of numbers written with the digits 0, 1, 2, 3, 4, 5, 6, 7, 8, and 9.

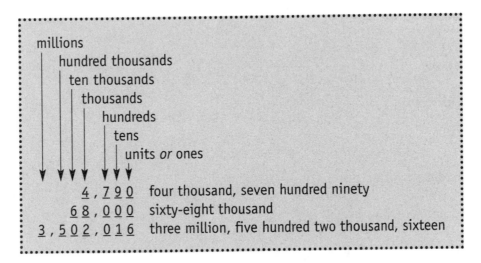

In 4,790 notice the zero. The zero holds the units or ones place. Notice also that commas separate the places in groups of three starting from the right. The number 4,790 is the correct way to write *four thousand seven hundred ninety.*

The number 68,000 has three zeros. The zeros hold the hundreds, tens, and units places.

The number 3,502,016 has two zeros. The zero between the 5 and the 2 holds the ten thousands place. The zero between the 2 and the 1 holds the hundreds place.

For problems 5 to 10 circle the number that is written correctly.

5. five thousand six hundred   5,060   5,600   5,006

6. seven thousand two   7,200   7,020   7,002

7. forty-four thousand, nine hundred   44,009   44,900   44,090

8. sixty-two thousand, four hundred three   62,403   60,243   62,430

9. eight hundred twenty thousand, seven hundred   827,000   820,007   820,700

10. two million, three hundred thousand   2,300,000   2,000,300   2,003,000

Write the following numbers.

11. forty-nine thousand, seven hundred thirty-six

12. four thousand eighty-two

13. two hundred four thousand, nine hundred seventy-two

14. three hundred thirty-nine

15. five thousand, one hundred nine

16. thirteen thousand

17. eight hundred two

18. thirty-four thousand, eighty-six

19. six hundred eighty thousand

20. two hundred forty thousand, seven hundred

21. one million, three hundred forty thousand, six hundred

22. three million, four hundred twenty-five thousand, one hundred

23. nine million, eight hundred seventy-three thousand

24. three hundred thirty-three thousand, three hundred

Rounding Whole Numbers

A **round number**—a number that ends with zeros—is easy to work with. Later, when you learn to estimate answers, you will often round the numbers in problems to similar numbers that end with zeros.

Is the number 26 closer to 20 or to 30? The number 26 is closer to 30 than to 20. The number 26 rounded to the nearest ten is 30.

To round a whole number, do the following steps.

STEP 1   Underline the digit in the place you are rounding to.

STEP 2   **a.** If the digit to the right of the underlined digit is *greater than or equal to 5,* add 1 to the underlined digit.

b. If the digit to the right of the underlined digit is *less than 5,* leave the underlined digit as it is.

STEP 3   Change all the digits to the right of the underlined digit to zeros.

EXAMPLE 1   Round 479 to the nearest ten.

STEP 1   Underline 7, the digit in the tens place. $4\underline{7}9 \rightarrow 480$

STEP 2   The digit to the right of 7 is 9. Since 9 is greater than 5, add 1 to 7. $7 + 1 = 8$

STEP 3   Change the digit to the right of 8 to 0.

EXAMPLE 2   Round 4,316 to the nearest hundred.

STEP 1   Underline 3, the digit in the hundreds place. $4,\underline{3}16 \rightarrow 4,300$

STEP 2   The digit to the right of 3 is 1. Since 1 is less than 5, leave 3 as it is.

STEP 3   Change the digits to the right of 3 to zeros.

Round each number to the nearest ten.

1. 76

2. 128

3. 261

4. 5,069

5. 288

6. 92

7. 14,355

8. 334

9. 983

Round each number to the nearest hundred.

10. 294	**11.** 3,528	**12.** 661
13. 2,319	**14.** 453	**15.** 18,766
16. 927	**17.** 6,859	**18.** 371

Round each number to the nearest thousand.

19. 3,614	**20.** 8,197	**21.** 12,543
22. 8,469	**23.** 235,520	**24.** 9,230
25. 26,612	**26.** 1,199	**27.** 62,903

Round each number to the nearest ten thousand.

28. 14,500	**29.** 26,125	**30.** 142,890
31. 81,777	**32.** 249,000	**33.** 173,208
34. 526,400	**35.** 78,312	**36.** 1,237,000

37. Sam and Serena bought a new car for $12,942. Which of the following rounds the price of the car to the nearest thousand dollars?

    a. $12,940
    b. $12,900
    c. $13,000
    d. $10,000

38. A real estate appraiser told Mrs. Vega that the value of her house was $136,500. Which of the following expresses the value of the house to the nearest ten thousand dollars?

    a. $136,500
    b. $137,000
    c. $140,000
    d. $100,000

39. Cal Hamilton ran for state assembly in his district. He won with a total of 213,802 votes. Which of the following expresses the number of votes to the nearest ten thousand?

    a. 200,000
    b. 210,000
    c. 213,800
    d. 214,000

ADDITION

Addition Skills Inventory

Do all the problems that you can. There is no time limit. Work carefully and check your answers, but do not use outside help.

1. 4,825
    + 2,164

2. 51,937
    + 35,042

3. 210,976
    + 384,012

4. 23
    54
    + 69

5. 72
    83
    46
    + 54

6. 28
    50
    63
    48
    + 52

7. 927
    48
    413
    + 52

8. 46
    519
    53
    + 246

9. 506
    49
    732
    + 88

10. $41 + 638 + 23 =$

11. $215 + 23 + 44 + 7 =$

12. $8,126 + 75,634 + 29 =$

13. $52 + 3,497 + 8 + 21,046 =$

14. $476,279 + 861,557 =$

15. $16 + 38 + 4 + 125 + 2 =$

16. Round each number to the nearest *hundred* and add.

    $6,954 + 1,326 + 4,579$

17. Round each number to the nearest *thousand* and add.

    $29,847 + 31,866 + 49,230$

18. Round each number to the nearest *thousand* and add.

    59,146 + 28,759 + 61,238 + 49,852

19. Round each number to the nearest *hundred* and add.

    5,418 + 6,059 + 7,837 + 2,364

20. During a basketball season, the Atlanta Hawks won 56 games and lost 26. What total number of games did they play?

21. In an election for governor of Wyoming, the winner received 97,299 votes. The second-place candidate received 70,661 votes. The third-place candidate received 6,897 votes. Find the combined number of votes for the three leading candidates.

22. The attendance for 4 days at a county fair were 6,380 the first day, 5,963 the second day, 6,754 the third day, and 7,018 the last day. Round each day's attendance to the nearest hundred. Then find an estimate of the total attendance using the rounded numbers.

23. The following deductions were taken from Mr. Munro's paycheck: $15.86 for federal tax, $26.46 for social security, $8.36 for state tax, and $3.81 for city tax. Find the total amount of the deductions from Mr. Munro's check.

Addition Skills Inventory Chart

If you missed more than one problem on any group below, work through the practice pages for that group. Then redo the problems you got wrong on the Addition Skills Inventory. If you had a passing score on all five groups of problems, redo any problem you missed and begin the Subtraction Skills Inventory on page 33.

PROBLEM NUMBERS	SKILL AREA	PRACTICE PAGES
1, 2, 3	addition facts	15–18
4, 5, 6, 7, 8, 9	adding and carrying	19–22
10, 11, 12, 13, 14, 15	adding horizontally	23
16, 17, 18, 19	rounding and estimating	25–26
20, 21, 22, 23	applying addition	27–30

Basic Addition Facts

This page and the following page will help you learn the basic addition facts. These are the first examples of **mental math** in this book. Mental math refers to the problems you can do without pencil and paper. The addition facts are basic building blocks for further study of mathematics. You must *memorize* these facts. Check your answers to this exercise. Practice the facts you got wrong. Then try these problems again until you can get all the answers quickly and accurately.

Parts of an Addition Problem

$$\begin{array}{r} 4 \ \longleftarrow \text{addend} \\ + 5 \ \longleftarrow \text{addend} \\ \hline 9 \ \longleftarrow \text{sum or total} \end{array}$$

 Add.

1.
$\begin{array}{r} 7 \\ +2 \\ \hline \end{array}$
$\begin{array}{r} 4 \\ +4 \\ \hline \end{array}$
$\begin{array}{r} 6 \\ +5 \\ \hline \end{array}$
$\begin{array}{r} 2 \\ +3 \\ \hline \end{array}$
$\begin{array}{r} 7 \\ +1 \\ \hline \end{array}$
$\begin{array}{r} 2 \\ +4 \\ \hline \end{array}$
$\begin{array}{r} 5 \\ +0 \\ \hline \end{array}$
$\begin{array}{r} 3 \\ +9 \\ \hline \end{array}$
$\begin{array}{r} 5 \\ +8 \\ \hline \end{array}$
$\begin{array}{r} 0 \\ +9 \\ \hline \end{array}$

2.
$\begin{array}{r} 3 \\ +6 \\ \hline \end{array}$
$\begin{array}{r} 7 \\ +4 \\ \hline \end{array}$
$\begin{array}{r} 5 \\ +5 \\ \hline \end{array}$
$\begin{array}{r} 6 \\ +4 \\ \hline \end{array}$
$\begin{array}{r} 9 \\ +7 \\ \hline \end{array}$
$\begin{array}{r} 2 \\ +0 \\ \hline \end{array}$
$\begin{array}{r} 3 \\ +3 \\ \hline \end{array}$
$\begin{array}{r} 2 \\ +7 \\ \hline \end{array}$
$\begin{array}{r} 5 \\ +9 \\ \hline \end{array}$
$\begin{array}{r} 1 \\ +4 \\ \hline \end{array}$

3.
$\begin{array}{r} 9 \\ +2 \\ \hline \end{array}$
$\begin{array}{r} 0 \\ +4 \\ \hline \end{array}$
$\begin{array}{r} 1 \\ +6 \\ \hline \end{array}$
$\begin{array}{r} 7 \\ +6 \\ \hline \end{array}$
$\begin{array}{r} 8 \\ +1 \\ \hline \end{array}$
$\begin{array}{r} 7 \\ +5 \\ \hline \end{array}$
$\begin{array}{r} 5 \\ +2 \\ \hline \end{array}$
$\begin{array}{r} 0 \\ +3 \\ \hline \end{array}$
$\begin{array}{r} 6 \\ +9 \\ \hline \end{array}$
$\begin{array}{r} 3 \\ +4 \\ \hline \end{array}$

4.
$\begin{array}{r} 0 \\ +6 \\ \hline \end{array}$
$\begin{array}{r} 1 \\ +5 \\ \hline \end{array}$
$\begin{array}{r} 7 \\ +7 \\ \hline \end{array}$
$\begin{array}{r} 4 \\ +5 \\ \hline \end{array}$
$\begin{array}{r} 9 \\ +0 \\ \hline \end{array}$
$\begin{array}{r} 4 \\ +1 \\ \hline \end{array}$
$\begin{array}{r} 3 \\ +7 \\ \hline \end{array}$
$\begin{array}{r} 6 \\ +2 \\ \hline \end{array}$
$\begin{array}{r} 1 \\ +0 \\ \hline \end{array}$
$\begin{array}{r} 2 \\ +6 \\ \hline \end{array}$

5.
$\begin{array}{r} 4 \\ +8 \\ \hline \end{array}$
$\begin{array}{r} 2 \\ +5 \\ \hline \end{array}$
$\begin{array}{r} 1 \\ +7 \\ \hline \end{array}$
$\begin{array}{r} 8 \\ +8 \\ \hline \end{array}$
$\begin{array}{r} 8 \\ +3 \\ \hline \end{array}$
$\begin{array}{r} 5 \\ +6 \\ \hline \end{array}$
$\begin{array}{r} 6 \\ +0 \\ \hline \end{array}$
$\begin{array}{r} 7 \\ +3 \\ \hline \end{array}$
$\begin{array}{r} 4 \\ +9 \\ \hline \end{array}$
$\begin{array}{r} 6 \\ +8 \\ \hline \end{array}$

6.
$\begin{array}{r} 2 \\ +2 \\ \hline \end{array}$
$\begin{array}{r} 9 \\ +8 \\ \hline \end{array}$
$\begin{array}{r} 3 \\ +5 \\ \hline \end{array}$
$\begin{array}{r} 4 \\ +6 \\ \hline \end{array}$
$\begin{array}{r} 0 \\ +7 \\ \hline \end{array}$
$\begin{array}{r} 3 \\ +2 \\ \hline \end{array}$
$\begin{array}{r} 5 \\ +3 \\ \hline \end{array}$
$\begin{array}{r} 4 \\ +0 \\ \hline \end{array}$
$\begin{array}{r} 1 \\ +3 \\ \hline \end{array}$
$\begin{array}{r} 8 \\ +2 \\ \hline \end{array}$

7.
$\begin{array}{r} 8 \\ +0 \\ \hline \end{array}$
$\begin{array}{r} 5 \\ +7 \\ \hline \end{array}$
$\begin{array}{r} 9 \\ +1 \\ \hline \end{array}$
$\begin{array}{r} 9 \\ +3 \\ \hline \end{array}$
$\begin{array}{r} 1 \\ +2 \\ \hline \end{array}$
$\begin{array}{r} 8 \\ +5 \\ \hline \end{array}$
$\begin{array}{r} 4 \\ +3 \\ \hline \end{array}$
$\begin{array}{r} 8 \\ +6 \\ \hline \end{array}$
$\begin{array}{r} 3 \\ +0 \\ \hline \end{array}$
$\begin{array}{r} 1 \\ +1 \\ \hline \end{array}$

8.
$\begin{array}{r} 4 \\ +7 \\ \hline \end{array}$
$\begin{array}{r} 0 \\ +5 \\ \hline \end{array}$
$\begin{array}{r} 8 \\ +4 \\ \hline \end{array}$
$\begin{array}{r} 6 \\ +3 \\ \hline \end{array}$
$\begin{array}{r} 7 \\ +8 \\ \hline \end{array}$
$\begin{array}{r} 4 \\ +2 \\ \hline \end{array}$
$\begin{array}{r} 5 \\ +1 \\ \hline \end{array}$
$\begin{array}{r} 7 \\ +0 \\ \hline \end{array}$
$\begin{array}{r} 9 \\ +5 \\ \hline \end{array}$
$\begin{array}{r} 0 \\ +2 \\ \hline \end{array}$

9.

1	9	6	0	1	5	2	9	6	3
+8	+4	+7	+1	+9	+4	+8	+9	+1	+8

10.

6	2	8	2	8	0	9	3	0	7
+6	+1	+9	+9	+7	+8	+6	+1	+0	+9

11.

2	2	7	4	5	3	8	4	5	0
+4	+0	+5	+1	+6	+2	+5	+2	+4	+8

12.

6	5	1	7	1	3	9	8	6	8
+5	+5	+6	+7	+7	+5	+1	+4	+7	+9

13.

5	3	5	3	6	4	5	2	5	9
+0	+3	+2	+7	+0	+3	+1	+8	+3	+6

14.

7	3	9	4	2	8	4	1	6	2
+2	+6	+2	+8	+2	+0	+7	+8	+6	+3

15.

6	7	0	4	8	4	9	6	0	2
+4	+6	+6	+5	+8	+6	+3	+3	+1	+9

16.

0	1	3	2	6	8	1	0	3	7
+9	+4	+4	+6	+8	+2	+1	+2	+8	+9

17.

3	2	0	7	4	8	7	6	9	3
+9	+7	+3	+3	+0	+6	+0	+2	+9	+1

18.

4	7	1	2	9	5	0	9	2	0
+4	+4	+5	+5	+8	+7	+5	+4	+1	+4

19.

7	9	8	9	8	0	1	7	1	8
+1	+7	+1	+0	+3	+7	+2	+8	+9	+7

20.

5	5	6	1	4	1	3	9	6	0
+8	+9	+9	+0	+9	+3	+0	+5	+1	+0

Adding Larger Numbers

If you know the basic addition facts on pages 15 and 16, you are ready
to add larger numbers. Add the column at the right first, and then move
to the next column to the left. Continue until you have added each
column of figures.

To check an addition problem, you can add the numbers in each
column from the bottom.

EXAMPLE   35   **STEP 1** 5 + 2 = 7   CHECK   35   **STEP 1** 2 + 5 = 7
          + 62 **STEP 2** 3 + 6 = 9   + 62   **STEP 2** 6 + 3 = 9
          ----   ----
          97   97

Add and check.

1. 47   50   84   63   78   42   51   70
    + 42   + 38   + 15   + 24   + 10   + 56   + 22   + 19

2. 61   12   81   75   52   30   69   34
    + 15   + 76   + 17   + 22   + 35   + 46   + 20   + 52

3. 603   577   458   413   762   210   307
    + 285   + 321   + 201   + 564   + 125   + 648   + 652

4. 805   761   432   854   386   328   415
    + 163   + 135   + 245   + 133   + 402   + 560   + 383

5. 4,825   1,756   6,073   4,261   2,352   7,084
    + 3,054 + 6,213 + 2,515 + 3,428 + 6,043 + 2,713

6. 18,946   70,128   54,223   38,684   76,521
    + 51,032 + 16,641 + 32,564 + 20,314 + 13,054

7. 36,745   56,092   72,033   27,514   30,649
    + 42,031 + 23,805 + 21,563 + 22,384 + 54,320

8. 662,184   729,832   229,056   367,455
    + 234,013   + 240,144   + 500,832   + 422,301

9.

81	346	45	3,720	51,084	3,817
+ 15	+ 252	+ 42	+ 4,256	+ 27,505	+ 4,162

10.

408	69,043	27	523	861,042	56
+ 241	+ 20,516	+ 40	+ 271	+ 121,446	+ 22

11.

7,244	8,563	5,042	7,136	3,574	5,546
+ 2,351	+ 1,234	+ 2,635	+ 1,042	+ 3,103	+ 2,342

12.

8,027	3,329	6,203	4,815	8,047	1,966
+ 1,932	+ 4,060	+ 2,351	+ 2,114	+ 1,832	+ 4,032

13.

6,001	2,347	7,580	4,116	7,038	2,413
+ 2,573	+ 2,412	+ 1,316	+ 4,572	+ 2,521	+ 4,334

14.

405	216	537	640	885	235	716
+ 561	+ 423	+ 252	+ 118	+ 102	+ 452	+ 263

15.

24,207	54,156	32,854	25,401	69,821
+ 15,072	+ 30,422	+ 43,104	+ 22,367	+ 20,104

16.

946	415	723	302	729	655	482
+ 32	+ 61	+ 75	+ 83	+ 50	+ 43	+ 16

17.

53	13	62	47	56	34	80
+ 436	+ 206	+ 914	+ 532	+ 811	+ 205	+ 709

18.

3,224	7,152	8,210	9,348	2,416	3,468
+ 641	+ 807	+ 326	+ 341	+ 502	+ 330

19.

547	296	638	513	704	652
+ 6,151	+ 4,203	+ 5,141	+ 6,256	+ 8,235	+ 3,144

20.

52,364	35,086	20,316	27,290	54,763
+ 2,305	+ 1,213	+ 8,271	+ 2,607	+ 4,134

Adding and Carrying

When the sum of the numbers in a column has a 2-digit answer (such as 13 in Step 1 in the example below), write the digit on the right under the column you just added and carry the left digit to the next column.

Writing a digit in the next column is sometimes called **regrouping** or **renaming.** It is also called **carrying.** The idea is to add units with units, tens with tens, hundreds with hundreds, and so on.

EXAMPLE

$$\begin{array}{r} {\scriptstyle 1\ 1} \\ 546 \\ +\ 297 \\ \hline 843 \end{array}$$

STEP 1 6 + 7 = 13. Write 3 in the units column and 1 above the tens column.

STEP 2 1 + 4 + 9 = 14. Write 4 in the tens column and 1 above the hundreds column.

STEP 3 1 + 5 + 2 = 8

To check an addition problem, you can add from the bottom as you learned on page 17. You can also try the following method.

			Write down
$\begin{array}{r}546 \\ +297 \\ \hline 843\end{array}$	**STEP 1**	Add the first column on the right in your head and write the *entire* total. 6 + 7 = 13	13
	STEP 2	Add the next column to the left and write that *entire* total under the first total and one column to the left. 4 + 9 + 13	13 13
	STEP 3	Add the next column to the left and write that *entire* total under the second total and one more column to the left. 5 + 2 = 7	13 13 7
	STEP 4	Add the three totals.	843

This method can be used with any number of figures. Always remember to write the total of each column under the previous column *and* one column to the left.

Add and check. The first problem has been done for you.

1.
$$
\begin{array}{r} ^1 \\ 46 \\ +47 \\ \hline 93 \end{array}
\quad
\begin{array}{r} 13 \\ 8 \\ \hline 93 \end{array}
\qquad
\begin{array}{r} 57 \\ +83 \\ \hline \end{array}
\qquad
\begin{array}{r} 67 \\ +54 \\ \hline \end{array}
\qquad
\begin{array}{r} 43 \\ +29 \\ \hline \end{array}
\qquad
\begin{array}{r} 82 \\ +68 \\ \hline \end{array}
\qquad
\begin{array}{r} 29 \\ +41 \\ \hline \end{array}
$$

2.
$$
\begin{array}{r} 38 \\ +93 \\ \hline \end{array}
\qquad
\begin{array}{r} 63 \\ +29 \\ \hline \end{array}
\qquad
\begin{array}{r} 22 \\ +78 \\ \hline \end{array}
\qquad
\begin{array}{r} 54 \\ +99 \\ \hline \end{array}
\qquad
\begin{array}{r} 48 \\ +53 \\ \hline \end{array}
\qquad
\begin{array}{r} 73 \\ +89 \\ \hline \end{array}
$$

3.
$$
\begin{array}{r} 56 \\ +90 \\ \hline \end{array}
\qquad
\begin{array}{r} 62 \\ +59 \\ \hline \end{array}
\qquad
\begin{array}{r} 94 \\ +27 \\ \hline \end{array}
\qquad
\begin{array}{r} 17 \\ +55 \\ \hline \end{array}
\qquad
\begin{array}{r} 652 \\ +478 \\ \hline \end{array}
\qquad
\begin{array}{r} 409 \\ +583 \\ \hline \end{array}
$$

4.
$$
\begin{array}{r} 683 \\ +417 \\ \hline \end{array}
\qquad
\begin{array}{r} 257 \\ +683 \\ \hline \end{array}
\qquad
\begin{array}{r} 594 \\ +417 \\ \hline \end{array}
\qquad
\begin{array}{r} 267 \\ +936 \\ \hline \end{array}
\qquad
\begin{array}{r} 607 \\ +406 \\ \hline \end{array}
\qquad
\begin{array}{r} 219 \\ +389 \\ \hline \end{array}
$$

5.
$$
\begin{array}{r} 740 \\ +893 \\ \hline \end{array}
\qquad
\begin{array}{r} 584 \\ +427 \\ \hline \end{array}
\qquad
\begin{array}{r} 637 \\ +905 \\ \hline \end{array}
\qquad
\begin{array}{r} 842 \\ +369 \\ \hline \end{array}
\qquad
\begin{array}{r} 715 \\ +486 \\ \hline \end{array}
\qquad
\begin{array}{r} 709 \\ +983 \\ \hline \end{array}
$$

6.
$$
\begin{array}{r} 384 \\ +429 \\ \hline \end{array}
\qquad
\begin{array}{r} 968 \\ +518 \\ \hline \end{array}
\qquad
\begin{array}{r} 314 \\ +98 \\ \hline \end{array}
\qquad
\begin{array}{r} 283 \\ +56 \\ \hline \end{array}
\qquad
\begin{array}{r} 607 \\ +95 \\ \hline \end{array}
\qquad
\begin{array}{r} 966 \\ +35 \\ \hline \end{array}
$$

7.
$$
\begin{array}{r} 748 \\ +87 \\ \hline \end{array}
\qquad
\begin{array}{r} 825 \\ +59 \\ \hline \end{array}
\qquad
\begin{array}{r} 346 \\ +64 \\ \hline \end{array}
\qquad
\begin{array}{r} 293 \\ +53 \\ \hline \end{array}
\qquad
\begin{array}{r} 53 \\ +469 \\ \hline \end{array}
\qquad
\begin{array}{r} 64 \\ +377 \\ \hline \end{array}
$$

8.
$$
\begin{array}{r} 68 \\ +580 \\ \hline \end{array}
\qquad
\begin{array}{r} 72 \\ +469 \\ \hline \end{array}
\qquad
\begin{array}{r} 93 \\ +708 \\ \hline \end{array}
\qquad
\begin{array}{r} 87 \\ +506 \\ \hline \end{array}
\qquad
\begin{array}{r} 41 \\ +289 \\ \hline \end{array}
\qquad
\begin{array}{r} 57 \\ +418 \\ \hline \end{array}
$$

9.
$$
\begin{array}{r} 1 \\ 30 \\ +9 \\ \hline \end{array}
\qquad
\begin{array}{r} 5 \\ 47 \\ +8 \\ \hline \end{array}
\qquad
\begin{array}{r} 2 \\ 81 \\ +5 \\ \hline \end{array}
\qquad
\begin{array}{r} 3 \\ 25 \\ +6 \\ \hline \end{array}
\qquad
\begin{array}{r} 2 \\ 56 \\ +5 \\ \hline \end{array}
\qquad
\begin{array}{r} 9 \\ 37 \\ +8 \\ \hline \end{array}
$$

10.	6	5	66	21	89	92
	52	83	81	49	67	10
	+ 9	+ 4	+ 56	+ 43	+ 50	+ 39

11.	87	60	31	79	20	95
	19	54	28	96	74	80
	+ 72	+ 38	+ 44	+ 83	+ 627	+ 417

12.	70	51	42	78	63	84
	57	91	93	12	85	43
	+ 202	+ 329	+ 516	+ 490	+ 854	+ 977

13.	367	236	413	146	243	880
	520	153	648	73	10	77
	+ 124	+ 875	+ 381	+ 718	+ 256	+ 523

14.	91	15	23	646	930	347
	31	75	22	60	15	38
	78	76	87	52	37	67
	+ 66	+ 31	+ 99	+ 944	+ 376	+ 421

15.	685	597	263	760	461	298
	691	283	161	218	919	709
	274	406	247	322	653	395
	+ 394	+ 938	+ 459	+ 938	+ 597	+ 471

16.	57	43	86	94	40	27
	35	51	65	21	83	40
	12	74	84	39	26	67
	83	32	46	45	29	28
	+ 58	+ 36	+ 97	+ 80	+ 33	+ 86

17.

8,779	4,855	7,630	4,596
2,286	2,849	4,108	8,892
+ 5,269	+ 1,754	+ 7,068	+ 4,625

18.

3,948	6,787	4,108	9,081
7,758	3,316	7,915	8,752
6,799	4,213	3,736	2,978
+ 2,437	+ 5,449	+ 2,615	+ 7,093

19.

76,493	44,538	27,881
66,590	64,908	92,855
+ 27,286	+ 70,435	+ 33,064

20.

518	114	806	638	441
782	726	992	793	348
764	953	528	136	635
207	199	666	483	914
+ 843	+ 727	+ 894	+ 397	+ 679

21.

520	714	863	933
2,186	7,465	6,097	9,487
706	302	141	761
5,807	4,374	8,262	3,309
+ 935	+ 491	+ 516	+ 743

22.

596,868	998,273	610,199
243,117	289,713	583,045
646,245	665,014	236,980
+ 695,048	+ 959,084	+ 346,706

Adding Numbers Written Horizontally

When the numbers you want to add are not in vertical columns, rewrite them so that the units are under the units, the tens are under the tens, and so on. Always line up the units column *first*.

EXAMPLE $25 + 6 + 423 =$ **REWRITE AS**
$$\begin{array}{r} 25 \\ 6 \\ + 423 \\ \hline 454 \end{array}$$

Add and check.

1. $93 + 55 + 34 =$ $23 + 18 + 96 =$

2. $607 + 12 + 344 =$ $58 + 752 + 29 =$

3. $53 + 618 + 9 + 47 =$ $7 + 24 + 806 + 63 =$

4. $982 + 3,507 + 73 + 184 =$ $26 + 745 + 66 + 4,329 =$

5. $8,196 + 75,883 + 29 + 334 =$ $232 + 80,465 + 19 + 1,591 =$

6. $8,779 + 2,286 + 5,269 =$ $4,855 + 2,849 + 1,754 =$

7. $7,630 + 4,108 + 7,068 =$ $4,596 + 8,892 + 4,625 =$

Addition Shortcuts

Add each problem in your head.

1. $42 + 8 =$ $23 + 7 =$ $3 + 87 =$

2. $76 + 4 =$ $1 + 59 =$ $2 + 68 =$

3. $9 + 31 =$ $6 + 14 =$ $65 + 5 =$

The answer to each of the previous problems ends with zero. Two numbers whose sum ends in zero are sometimes called **compatible pairs.**

Look at the example below. You start to add $13 + 9 = 22$ but notice that $9 + 11 = 20$. Add these numbers first. Remember that *the numbers in an addition problem may be added in any order.* Look for a compatible pair, and add the pair first.

EXAMPLE $13 + 9 + 11 =$ The numbers 9 and 11 are a compatible pair. $9 + 11 = 20$

         $13 + \quad 20 \quad = 33$ Rewrite the problem as the sum of $13 + 20$.

Find a compatible pair. Rewrite the problem. Then add.

4. $62 + 8 + 5 =$ $18 + 45 + 5 =$

5. $19 + 13 + 7 =$ $6 + 94 + 13 =$

6. $36 + 12 + 4 =$ $52 + 8 + 21 =$

7. $9 + 14 + 51 =$ $19 + 27 + 3 =$

Find two compatible pairs. Rewrite the problem. Then add.

8. $28 + 17 + 2 + 3 + 9 =$ $12 + 9 + 6 + 21 + 8 =$

9. $6 + 35 + 11 + 14 + 5 =$ $4 + 35 + 13 + 26 + 5 =$

10. $15 + 19 + 1 + 27 + 3 =$ $21 + 14 + 42 + 8 + 6 =$

11. $43 + 7 + 8 + 62 + 5 =$ $17 + 11 + 5 + 9 + 3 =$

Rounding, Estimating, and Using a Calculator

Review the rules for rounding numbers on page 11. When the digit to the right of the number you are rounding to is greater than or equal to 5, you must add 1 to the underlined digit. If the underlined digit is a 9, the digit to the left of 9 changes. Look at this example carefully.

EXAMPLE   Round 496 to the nearest ten. $4\underline{9}6 \rightarrow 500$

STEP 1   Underline 9, the digit in the tens place.

STEP 2   The digit to the right of 9 is 6. Since 6 is greater than 5, add 1 to 9. $9 + 1 = 10$
Write 0 in the tens column, and add 1 to the hundreds column. $1 + 4 = 5$

STEP 3   Change the digit in the units place to 0.

Notice, in the last example, that when you round 496 to the nearest ten, you get 500. When you round 496 to the nearest hundred, you also get 500.

Round each number to the nearest *ten*.

1. 296   98   3,095   649   4,583   12,487

Round each number to the nearest *hundred*.

2. 3,972   24,968   8,391   7,083   42,952   6,546

Round each number to the nearest *thousand*.

3. 49,806   75,928   219,655   8,962   69,506   2,384

Estimation means finding a reasonable answer. An estimate is not exact, but it is close. Rounding the numbers in a problem is a good way to estimate an answer.

Round each number to the nearest *hundred*. Then add the rounded numbers.

4. $3,948 + 758 + 6,799 + 437$ $9,081 + 752 + 2,978 + 93$

5. $787 + 3,316 + 213 + 5,449$ $520 + 2,186 + 706 + 5,807$

6. $108 + 7,915 + 3,736 + 615$ $714 + 7,465 + 302 + 4,974$

Round each number to the nearest *thousand*. Then add the rounded numbers.

7. 76,493 + 6,590 + 27,286   19,863 + 6,097 + 28,141

8. 4,538 + 64,908 + 70,435   7,933 + 9,487 + 16,761

9. 27,881 + 92,855 + 3,064   41,935 + 2,491 + 18,516

Using a Calculator   A calculator is a convenient tool for getting an exact answer.

EXAMPLE   Use a calculator to find the sum of 396 + 87 + 5.

Press the following keys on a calculator: $\boxed{3}\,\boxed{9}\,\boxed{6}\,\boxed{+}\,\boxed{8}\,\boxed{7}\,\boxed{+}\,\boxed{5}\,\boxed{=}$

The calculator display should read _____ 488.

For problems 10 to 12, round each number to the nearest *hundred* and find the sum of the rounded numbers. Then use a calculator to find the exact answers.

10. 518 + 782 + 764 + 207 + 843 =

11. 114 + 2,726 + 3,953 + 199 + 4,727 =

12. 6,806 + 992 + 5,528 + 4,666 + 894 =

For problems 13 to 15, round each number to the nearest *thousand* and find the sum of the rounded numbers. Then use a calculator to find the exact answers.

13. 596,868 + 43,117 + 646,245 + 95,048 =

14. 98,273 + 289,713 + 65,014 + 959,084 =

15. 610,199 + 83,045 + 36,980 + 346,706 =

Applying Your Addition Skills

On this page and the following three pages are problems that require you to apply your addition skills to practical problems. In each problem, pay close attention to the language that tells you to add. Watch for words such as **sum, total,** and **combined.**

In solving money problems, be sure to add dollars in the dollars columns and cents in the cents columns. For example, to find the sum of $4, $2.95, and 36¢, set the numbers under each other in this way:

$$
\begin{array}{r}
\$4.00 \\
2.95 \\
+ \quad .36 \\
\hline
\$7.31
\end{array}
$$

Solve and write the correct label, such as $, miles, or pounds, next to each answer.

1. On Monday Alphonse drove 12 miles to take his children to school, 18 miles to get to his job, 9 miles to get to the supermarket after work, and 15 miles to get back home. How far did he drive that day?

2. On Debbie's last telephone bill there was a monthly service charge of $8.27, long distance calls for $15.93, and tax for $2.10. What was her total telephone bill?

3. Attendance at a 4-day basketball tournament was the following: Thursday, 2,640; Friday, 2,873; Saturday, 2,892; and Sunday, 2,916. What was the total attendance at the tournament?

4. A group of friends started a business. Mr. Reilly put in $7,680, Mr. Flood put in $4,275, and Mr. Murphy put in $8,923. What is their combined investment?

5. When David sold his television to his sister, he got $85.00 for it. This was $47.50 less than what he paid for the television. How much did he pay for the television?

6. When the register in the Lake Pharmacy broke down, the cashier had to add up the day's receipts by hand. Purchases totaled $650.38 for prescription drugs, $578.12 for over-the-counter drugs, and $498.84 for cosmetics. What was the cashier's total?

7. In one year, the immigration into the United States from Europe was 91,183; from Asia it was 119,984; and from Africa it was 5,537. What was the combined immigration from these areas that year?

8. The weights of the children in the Wiley family are Mark, 56 pounds; Kim, 89 pounds; David, 90 pounds; Alice, 112 pounds; and Sara, 133 pounds. What is the sum of the weights of the Wiley children?

9. On Michelle's last telephone bill she had the following long-distance charges: a call to Middletown for $0.85, a call to Central City for $8.20, a call to Springfield for $4.80, and a call to Portland for $0.95. What was the total of her long-distance calls?

10. Pete loaded several boxes onto a freight elevator. Box A weighed 165 pounds, box B weighed 212 pounds, box C weighed 126 pounds, box D weighed 197 pounds, box E weighed 135 pounds, box F weighed 220 pounds, box G weighed 118 pounds, and box H weighed 136 pounds. Find the combined weight of the boxes.

11. The freight elevator in the last problem can carry 1,500 pounds. Can the elevator safely carry all the boxes from the last problem?

Use the following information to answer questions 12 and 13.

The population of the boroughs of New York City are The Bronx, 1,187,984; Brooklyn, 2,240,384; Manhattan, 1,536,220; Queens, 1,975,676; and Staten Island, 402,372.

 12. Round the population of each borough to the nearest hundred thousand. Then find the estimate of the total population of New York City using the rounded numbers.

 13. Use a calculator to find the exact population of New York City.

14. In one year in the United States, the per capita (per person) purchase of meat and fish was 111 pounds of beef, 3 pounds of lamb, 2 pounds of veal, 62 pounds of pork, 12 pounds of fish, and 51 pounds of poultry. Find the total per capita purchase of these items.

15. Beverly wanted to see how much of her daily expense money went for taxes. This is what she found: $3.52 in tax to fill her gasoline tank, $1.08 in tax for her cigarettes, $0.48 tax on her lunch, and $2.34 tax on a sweater she bought on sale. Find the total amount Beverly paid in tax for these items.

16. When Carmen and Julio went out to eat, they had a pizza for $10.95, two salads for $5.90, and two sodas for $1.80. The tax on their bill was 93¢, and they left a tip of $2.80. Find the whole amount they spent for lunch.

17. The distance from Cincinnati to Columbus is 108 miles. The distance from Columbus to Cleveland is 133 miles. What is the distance from Cincinnati to Cleveland by way of Columbus?

18. In a recent year, the U.S. army had 81,291 officers and 413,991 enlisted soldiers. What was the combined number of officers and enlisted soldiers?

19. Al decided to go off his diet for one afternoon, so he ate his favorite foods. For a snack he had a ginger ale (75 calories) and pretzel sticks (200 calories). Later he ate a cheese pizza (555 calories) and drank a cola (95 calories). Find the total number of calories Al consumed that afternoon.

20. The weekly budget for the Thomas family is $90.25 for rent, $107.00 for food, $30.00 for utilities (gas, electricity, and telephone), $36.00 for clothing and laundry, $53.00 for entertainment, and $25.00 for savings. What is their total weekly budget?

Use the following information for questions 21 and 22.

The area of each of the five Great Lakes is as follows: Ontario, 7,540 square miles; Erie, 9,940 square miles; Michigan, 22,400 square miles; Huron, 23,010 square miles; and Superior, 31,820 square miles.

21. Which of the following is closest to the combined area in square miles of the Great lakes?

   a. 75,000
   b. 85,000
   c. 95,000
   d. 105,000

22. Use a calculator to find the exact combined area of the Great Lakes.

Addition Review

This review covers the first two chapters of this book.

1. What is the value of the digit 2 in the number 342,508?

2. What is 14,763 rounded to the nearest hundred?

3. What is $481,297 rounded to the nearest ten thousand?

4. 5,324
    + 2,145

5. 23,864
    + 51,133

6. 348,065
    + 251,822

7. 56
    87
    + 25

8. 63
    54
    97
    + 81

9. 32
    40
    76
    51
    + 49

10. 816
    53
    228
    + 46

11. 53
    226
    45
    + 513

12. 619
    58
    417
    + 96

13. $63 + 214 + 55 =$

14. $425 + 36 + 77 + 4 =$

15. $817 + 1,329 + 638 =$

16. $12 + 3,407 + 8 + 496 =$

17. $796,041 + 532,408 =$

18. $27 + 48 + 3 + 901 + 2 =$

19. Round each number to the nearest *hundred* and add.

    $3,972 + 1,258 + 3,417$

20. Round each number to the nearest *thousand* and add.

69,836 + 71,254 + 33,927

 21. The population of Baltimore is 746,014. Round the number to the nearest ten thousand.

 22. In a month Anne paid $375.00 for rent, $38.94 for gas and electricity, $65.47 for telephone, and $110.00 for her car payment. Find the total of these expenses.

 23. Uncle Bob's car dealership sold 483 cars the first year they were open. The second year they sold 238 more than the first year. Find the total sales for the second year.

 24. There are four movie theatres in Sunset City. The Mayflower seats 438, the Colony seats 261, the Pilgrim seats 397, and the Standish seats 564. Which of the following is closest to the total number of seats in the Sunset City movie theatres?

   a. 1,300 seats
   b. 1,500 seats
   c. 1,700 seats
   d. 1,900 seats

Addition Review Chart

If you missed more than one problem on any group below, review the practice pages for those problems. Then redo the problems you got wrong before going on to the Subtraction Skills Inventory. If you had a passing score, redo any problem you missed and begin the Subtraction Skills Inventory on page 33.

PROBLEM NUMBERS	SKILL AREA	PRACTICE PAGES
1, 2, 3	place value	6–12
4, 5, 6	addition facts	15–18
7, 8, 9, 10, 11, 12	adding and carrying	19–22
13, 14, 15, 16, 17, 18	adding horizontally	23
19, 20, 21	rounding and estimating	25–26
22, 23, 24	applying addition	27–30

SUBTRACTION

Subtraction Skills Inventory

Do all the problems that you can. There is no time limit. Work carefully and check your answers, but do not use outside help.

1. 74
    − 30

2. 58
    − 27

3. 542
    − 241

4. 759
    − 207

5. 47
    − 29

6. 764
    − 19

7. 9,183
    − 2,097

8. 4,167
    − 2,758

9. 402
    − 358

10. 500
     − 276

11. 3,008
     − 1,956

12. 10,000
     − 7,049

13. 528 − 39 =

14. 425 − 276 =

15. 23,532 − 6,548 =

16. 70,060 − 4,118 =

17. 110 − 64 =

18. 73 − 25 =

19. Round each number to the nearest *hundred* and subtract.

    7,498 − 2,379

20. Round each number to the nearest *thousand* and subtract.

59,614 − 29,058

21. Round each number to the nearest *ten thousand* and subtract.

485,311 − 296,347

22. Joe's gross salary is $413.25. His employer deducts $111.11 from Joe's paycheck for taxes and benefits. What is Joe's take-home pay?

23. Mr. Greenwald bought a chair for $86.95. How much change did he receive if he paid with a $100 bill?

24. Before the Walek family went on vacation, their odometer (mileage dial) had a reading of 34,925 miles. When they returned, it read 36,059 miles. How far did they drive on their vacation?

25. Clarktown has 3,595 registered voters. At the last election, 2,459 people voted in Clarktown. How many of the registered voters did not vote?

Subtraction Skills Inventory Chart

If you missed more than one problem on any group below, work through the practice pages for that group. Then redo the problems you got wrong on the Subtraction Skills Inventory. If you had a passing score on all seven groups of problems, redo any problem you missed and begin the Multiplication Skills Inventory on page 60.

PROBLEM NUMBERS	SKILL AREA	PRACTICE PAGES
1, 2, 3, 4	subtraction facts	35–38
5, 6, 7, 8	subtracting and borrowing	39–43
9, 10, 11, 12	subtracting from zeros	45–48
13, 14, 15, 16	subtracting horizontally	44, 48
17, 18	subtraction shortcuts	49
19, 20, 21	rounding and estimating	50–51
22, 23, 24, 25	applying subtraction	52–56

Basic Subtraction Facts

This page and the following page will help you learn the basic subtraction facts. These are more examples of **mental math.** The subtraction facts are basic building blocks for further study of mathematics. You must *memorize* these facts. Check your answers to this exercise. Practice the facts you got wrong. Then try these problems again until you can get all the answers quickly and accurately.

Parts of a Subtraction Problem

6 ◄——— minuend
− 2 ◄——— subtrahend
4 ◄——— difference

| 1. | 5 −3 | 7 −6 | 8 −3 | 13 −5 | 11 −3 | 7 −2 | 8 −0 | 11 −9 | 10 −7 | 9 −9 |

| 2. | 7 −1 | 5 −2 | 13 −7 | 14 −6 | 3 −2 | 4 −3 | 15 −6 | 10 −9 | 9 −3 | 2 −1 |

| 3. | 6 −4 | 5 −0 | 10 −2 | 13 −4 | 8 −5 | 17 −8 | 7 −4 | 3 −3 | 15 −9 | 11 −8 |

| 4. | 1 −1 | 14 −5 | 6 −3 | 8 −7 | 9 −6 | 15 −8 | 11 −6 | 10 −6 | 4 −4 | 12 −7 |

| 5. | 9 −2 | 3 −0 | 12 −5 | 8 −8 | 9 −4 | 10 −3 | 13 −6 | 7 −7 | 12 −8 | 8 −4 |

| 6. | 10 −1 | 14 −9 | 12 −6 | 6 −0 | 7 −5 | 11 −7 | 9 −5 | 10 −4 | 1 −0 | 8 −1 |

| 7. | 16 −7 | 8 −6 | 10 −5 | 5 −4 | 6 −2 | 5 −1 | 7 −3 | 14 −8 | 18 −9 | 11 −4 |

| 8. | 4 −0 | 8 −2 | 13 −8 | 9 −1 | 6 −6 | 12 −3 | 3 −1 | 15 −7 | 11 −5 | 4 −1 |

9.

9	2	14	16	7	11	5	13	2	12
−7	−2	−7	−9	−0	−2	−5	−9	−0	−4

10.

6	17	9	6	16	12	9	4	10	13
−5	−9	−8	−1	−8	−9	−0	−2	−8	−6

11.

8	7	10	5	3	10	6	13	7	11
−3	−2	−7	−2	−2	−9	−4	−4	−4	−8

12.

6	15	4	3	9	7	10	6	9	8
−3	−8	−4	−0	−4	−7	−1	−0	−5	−1

13.

10	5	18	8	6	15	9	16	5	12
−5	−1	−9	−2	−6	−7	−7	−9	−5	−4

14.

9	12	10	7	11	11	7	14	15	2
−8	−9	−8	−6	−3	−9	−1	−6	−6	−1

15.

10	17	15	14	9	10	9	8	13	8
−2	−8	−9	−5	−6	−6	−2	−8	−6	−4

16.

12	11	1	8	6	14	4	9	3	4
−6	−7	−0	−6	−2	−8	−0	−1	−1	−1

17.

14	11	2	17	16	4	5	13	8	9
−7	−2	−0	−9	−8	−2	−3	−5	−0	−9

18.

13	4	9	5	8	3	1	8	11	12
−7	−3	−3	−0	−5	−3	−1	−7	−6	−7

19.

12	10	12	14	7	10	16	5	7	11
−5	−3	−8	−9	−5	−4	−7	−4	−3	−4

20.

13	12	11	2	7	13	6	6	9	13
−8	−3	−5	−2	−0	−9	−5	−1	−0	−6

Subtracting Larger Numbers

If you know the basic subtraction facts on pages 35 and 36, you are ready to subtract larger numbers. Subtract the column at the right first, and then move to the next column to the left. Continue until you have subtracted each column of figures.

To check a subtraction problem, add the answer to the bottom number of the original problem. The sum should be the top number of the original problem.

EXAMPLE			CHECH	
46	**STEP 1** $6 - 5 = 1$		46	
-25	**STEP 2** $4 - 2 = 2$		-25	**STEP 1** $5 + 1 = 6$
21			21	**STEP 2** $2 + 2 = 4$
			46 ✓	

Subtract and check.

1.
$$\begin{array}{r} 34 \\ -22 \end{array} \quad \begin{array}{r} 69 \\ -35 \end{array} \quad \begin{array}{r} 38 \\ -31 \end{array} \quad \begin{array}{r} 52 \\ -32 \end{array} \quad \begin{array}{r} 75 \\ -51 \end{array} \quad \begin{array}{r} 81 \\ -40 \end{array} \quad \begin{array}{r} 62 \\ -11 \end{array} \quad \begin{array}{r} 79 \\ -58 \end{array}$$

2.
$$\begin{array}{r} 60 \\ -40 \end{array} \quad \begin{array}{r} 35 \\ -21 \end{array} \quad \begin{array}{r} 77 \\ -73 \end{array} \quad \begin{array}{r} 74 \\ -40 \end{array} \quad \begin{array}{r} 58 \\ -32 \end{array} \quad \begin{array}{r} 41 \\ -21 \end{array} \quad \begin{array}{r} 37 \\ -15 \end{array} \quad \begin{array}{r} 62 \\ -21 \end{array}$$

3.
$$\begin{array}{r} 364 \\ -263 \end{array} \quad \begin{array}{r} 751 \\ -401 \end{array} \quad \begin{array}{r} 523 \\ -321 \end{array} \quad \begin{array}{r} 952 \\ -940 \end{array} \quad \begin{array}{r} 540 \\ -230 \end{array} \quad \begin{array}{r} 686 \\ -251 \end{array} \quad \begin{array}{r} 739 \\ -628 \end{array}$$

4.
$$\begin{array}{r} 249 \\ -134 \end{array} \quad \begin{array}{r} 553 \\ -243 \end{array} \quad \begin{array}{r} 916 \\ -503 \end{array} \quad \begin{array}{r} 692 \\ -491 \end{array} \quad \begin{array}{r} 485 \\ -324 \end{array} \quad \begin{array}{r} 727 \\ -512 \end{array} \quad \begin{array}{r} 609 \\ -203 \end{array}$$

5.
$$\begin{array}{r} 650 \\ -440 \end{array} \quad \begin{array}{r} 848 \\ -232 \end{array} \quad \begin{array}{r} 567 \\ -356 \end{array} \quad \begin{array}{r} 683 \\ -421 \end{array} \quad \begin{array}{r} 944 \\ -703 \end{array} \quad \begin{array}{r} 871 \\ -260 \end{array} \quad \begin{array}{r} 957 \\ -342 \end{array}$$

6.
$$\begin{array}{r} 887 \\ -352 \end{array} \quad \begin{array}{r} 543 \\ -243 \end{array} \quad \begin{array}{r} 781 \\ -300 \end{array} \quad \begin{array}{r} 942 \\ -611 \end{array} \quad \begin{array}{r} 889 \\ -278 \end{array} \quad \begin{array}{r} 468 \\ -432 \end{array} \quad \begin{array}{r} 982 \\ -581 \end{array}$$

7.

772	908	197	974	567	534	894
-341	-305	-122	-923	-524	-311	-423

8.

8,559	7,388	9,161	9,798	9,897	7,258
$-4,208$	$-6,234$	$-5,031$	$-9,624$	$-3,572$	$-6,125$

9.

6,235	7,819	6,798	4,272	5,467	7,948
$-5,221$	$-4,308$	$-6,217$	$-4,122$	$-3,245$	$-2,136$

10.

7,245	4,278	2,657	6,867	7,886	7,980
$-7,123$	$-3,072$	$-1,452$	$-3,260$	$-4,253$	$-3,210$

11.

91,527	87,496	47,086	53,484	68,237
$-20,415$	$-32,422$	$-17,025$	$-33,251$	$-32,025$

12.

86,947	56,739	92,465	75,826	59,428
$-32,825$	$-54,231$	$-22,143$	$-23,521$	$-41,314$

13.

688,540	628,345	923,457	421,276	477,563
$-287,430$	$-125,214$	$-902,133$	$-311,043$	$-251,412$

14.

787,445	255,694	860,956	568,798	874,395
$-257,312$	$-241,454$	$-360,224$	$-325,416$	$-211,243$

Subtracting and Borrowing

....

When a digit in the subtrahend (the bottom number) is too large to subtract from the digit above it, you have to **borrow** from the next column to the left in the minuend (the top number). Borrowing is sometimes called **regrouping** or **renaming.**

EXAMPLE $^{8}\cancel{9}^{1}3$

       $-\ 16$

       $\overline{\ 77}$

STEP 1 Since you cannot subtract 6 from 3, borrow 1 ten from the tens column. (10 + 3 = 13)

Put a small 1 next to the 3 in the top number to show that it is now 13.

STEP 2 Cross out the 9 in the tens column and make it an 8 to show that you have borrowed 1 ten.

STEP 3 Subtract the units 13 − 6 = 7.

STEP 4 Subtract the tens 8 − 1 = 7.

CHECK $\left.\begin{array}{r} 93 \\ -\ 16 \\ \hline 77 \end{array}\right\}+$

       $\overline{\ 93}$ ✓   **STEP 5**   Check 16 + 77 = 93.

....

Subtract and check.

1. 67	95	48	68	84	76	47	94
− 8	− 9	− 9	− 9	− 7	− 7	− 8	− 8

2. 82	27	21	95	63	91	62	33
− 9	− 8	− 7	− 8	− 5	− 6	− 9	− 4

3. 83	95	50	68	97	52	45	64
− 25	− 39	− 28	− 49	− 58	− 27	− 16	− 26

4.

53	28	97	85	44	68	70	42
-36	-19	-28	-16	-27	-59	-32	-18

5.

57	61	92	36	48	51	25	74
-29	-34	-83	-18	-29	-36	-18	-27

6.

763	478	532	514	261	188	344
$-\ 9$	$-\ 9$	$-\ 8$	$-\ 6$	$-\ 4$	$-\ 9$	$-\ 7$

7.

251	742	927	381	793	640	365
$-\ 38$	$-\ 27$	$-\ 18$	$-\ 53$	$-\ 86$	$-\ 27$	$-\ 48$

8.

896	692	546	695	588	482	981
$-\ 88$	$-\ 85$	$-\ 37$	$-\ 88$	$-\ 79$	$-\ 75$	$-\ 63$

9.

371	684	143	467	355	653	266
-269	-547	-128	-349	-237	-409	-118

10.

458	691	854	794	652	498	385
-349	-227	-536	-478	-239	-379	-166

11.

278	445	764	893	972	331	597
-119	-108	-545	-655	-408	-125	-408

Sometimes you have to borrow more than once in the same problem. Many of the problems on this page and the following two pages will give you practice with this kind of subtraction problem.

EXAMPLE

$$\begin{array}{r} 2\,2\,3 \\ -1\,7\,8 \\ \hline 4\,5 \\ \hline 2\,2\,3 \checkmark \end{array}$$

$$\begin{array}{r} 4\,3\,,2\,4\,1 \\ -2\,8\,,9\,8\,5 \\ \hline 1\,4\,,2\,5\,6 \\ \hline 4\,3\,,2\,4\,1 \checkmark \end{array}$$

....

Subtract and check.

1.
$$\begin{array}{r} 795 \\ -\ 29 \end{array}$$
$$\begin{array}{r} 932 \\ -\ 38 \end{array}$$
$$\begin{array}{r} 866 \\ -\ 47 \end{array}$$
$$\begin{array}{r} 357 \\ -\ 28 \end{array}$$
$$\begin{array}{r} 614 \\ -\ 17 \end{array}$$
$$\begin{array}{r} 288 \\ -\ 59 \end{array}$$

2.
$$\begin{array}{r} 968 \\ -\ 74 \end{array}$$
$$\begin{array}{r} 141 \\ -\ 80 \end{array}$$
$$\begin{array}{r} 572 \\ -\ 91 \end{array}$$
$$\begin{array}{r} 543 \\ -\ 62 \end{array}$$
$$\begin{array}{r} 519 \\ -\ 53 \end{array}$$
$$\begin{array}{r} 328 \\ -\ 86 \end{array}$$

3.
$$\begin{array}{r} 876 \\ -\ 89 \end{array}$$
$$\begin{array}{r} 553 \\ -\ 69 \end{array}$$
$$\begin{array}{r} 417 \\ -\ 58 \end{array}$$
$$\begin{array}{r} 864 \\ -\ 87 \end{array}$$
$$\begin{array}{r} 241 \\ -\ 48 \end{array}$$
$$\begin{array}{r} 752 \\ -\ 57 \end{array}$$

4.
$$\begin{array}{r} 525 \\ -247 \end{array}$$
$$\begin{array}{r} 287 \\ -198 \end{array}$$
$$\begin{array}{r} 216 \\ -158 \end{array}$$
$$\begin{array}{r} 466 \\ -388 \end{array}$$
$$\begin{array}{r} 373 \\ -199 \end{array}$$
$$\begin{array}{r} 983 \\ -585 \end{array}$$

5.
$$\begin{array}{r} 7,263 \\ -\ 185 \end{array}$$
$$\begin{array}{r} 4,775 \\ -\ 683 \end{array}$$
$$\begin{array}{r} 5,176 \\ -\ 438 \end{array}$$
$$\begin{array}{r} 8,240 \\ -\ 327 \end{array}$$
$$\begin{array}{r} 2,334 \\ -\ 516 \end{array}$$
$$\begin{array}{r} 1,615 \\ -\ 807 \end{array}$$

6.
3,633	4,125	2,961	5,648	6,345	1,827
− 794	− 636	− 972	− 649	− 258	− 918

7.
7,792	8,235	2,047	5,167	4,660	7,133
− 4,829	− 3,516	− 1,839	− 2,758	− 1,857	− 2,924

8.
9,113	2,474	3,323	9,457	4,375	5,126
− 2,058	− 1,185	− 2,196	− 3,379	− 2,088	− 3,048

9.
7,488	6,337	6,154	9,310	3,124	8,725
− 2,499	− 3,878	− 2,165	− 4,975	− 2,368	− 4,956

10.
2,328	5,181	8,760	7,395	7,227	5,684
− 1,489	− 2,397	− 5,764	− 2,498	− 4,285	− 2,993

11.
8,233	6,145	5,608	1,875	7,620	4,144
− 7,148	− 4,139	− 2,786	− 1,096	− 4,586	− 2,139

12.
4,871	6,524	2,775	4,726	8,673	2,865
− 2,984	− 3,576	− 1,799	− 3,946	− 1,888	− 1,977

13.
96,973	42,763	75,823	87,214	59,656
− 4,982	− 5,589	− 6,458	− 6,209	− 5,458

14.
42,356	38,182	87,421	78,253	34,818
− 9,485	− 8,491	− 9,530	− 9,682	− 5,847

15.
92,309	69,762	86,583	47,757	81,912
− 8,958	− 4,974	− 5,999	− 6,858	− 1,943

16.
93,532	83,124	41,591	35,876	66,180
− 9,544	− 4,876	− 5,694	− 8,977	− 7,293

17.
74,925	53,286	42,755	54,278	45,188
− 23,936	− 43,689	− 22,697	− 32,489	− 21,279

18.
22,483	62,537	55,130	25,614	28,786
− 19,584	− 56,849	− 43,273	− 16,726	− 19,797

19.
678,054	327,583	227,150	732,122	934,662
− 97,347	− 58,946	− 82,366	− 51,475	− 66,539

20.
923,644	482,637	822,472	673,612	834,124
− 849,357	− 297,148	− 631,584	− 297,641	− 535,217

21.
543,950	437,032	144,480	738,259	685,978
− 418,513	− 214,877	− 142,895	− 589,767	− 498,369

Subtracting Numbers Written Horizontally

When the numbers you want to subtract are not in vertical columns, rewrite them with the larger number on top. Make sure that you line up the units under the units, the tens under the tens, and so on. Always line up the units column *first*.

EXAMPLE $7,522 - 971 =$ **REWRITE AS** $\begin{array}{r} 7,522 \\ -971 \\ \hline \end{array}$

Subtract and check.

1. $7,522 - 971 =$ $9,330 - 827 =$ $5,942 - 307 =$

2. $2,165 - 2,076 =$ $6,752 - 4,397 =$ $9,283 - 2,474 =$

3. $3,275 - 2,486 =$ $5,680 - 4,597 =$ $7,223 - 6,445 =$

4. $57,542 - 9,651 =$ $24,143 - 5,048 =$ $82,633 - 3,272 =$

5. $21,456 - 3,569 =$ $26,116 - 5,248 =$ $53,224 - 4,097 =$

6. $61,510 - 42,513 =$ $43,524 - 22,685 =$ $15,697 - 14,938 =$

Subtracting from Zeros

You cannot borrow from zero. When the digit in the column you want to borrow from is zero, move to the next column to the left that does *not* contain a zero. Read through the steps of the next example carefully.

EXAMPLE

$$\overset{3\ \ 1}{4\cancel{0}6}$$
$$-\ 159$$

STEP 1 You cannot subtract 9 from 6. You cannot borrow 1 from the zero in the tens column. Borrow 1 from the hundreds column and take it to the empty tens column. You now have 10 tens in the tens column. This leaves 3 hundreds.

$$\overset{3\ 9}{\underset{1\ 1}{4\cancel{0}6}}$$
$$-\ 159$$

STEP 2 Borrow 1 from the tens column and take it to the units column. You now have $10 + 6 = 16$ in the units column. This leaves 9 tens.

$$\overset{3\ 9}{\underset{1\ 1}{4\cancel{0}6}}$$
$$-\ 159$$
$$\overline{247}$$
$$\overline{406}\ ✓$$

STEP 3 Subtract the units $16 - 9 = 7$.
STEP 4 Subtract the tens $9 - 5 = 4$.
STEP 5 Subtract the hundreds $3 - 1 = 2$.
STEP 6 Check $159 + 247 = 406$.

Subtract and check.

1.
205	306	402	508	206	404
− 86	− 38	− 46	− 59	− 49	− 57

2.
602	501	803	902	401	607
− 58	− 29	− 37	− 44	− 53	− 69

3.
502	308	604	305	207	401
− 263	− 149	− 326	− 278	− 199	− 285

4.

300	400	600	200	900	700
− 37	− 45	− 63	− 34	− 51	− 28

5.

800	300	500	400	600	200
− 263	− 127	− 233	− 291	− 382	− 194

6.

2,040	3,060	1,050	4,080	3,070	5,020
− 463	− 392	− 465	− 387	− 991	− 438

7.

6,050	2,040	3,010	5,010	2,010	8,010
− 2,367	− 1,248	− 2,563	− 3,466	− 1,277	− 4,252

8.

3,004	5,006	2,007	4,003	9,002	6,001
− 973	− 425	− 347	− 621	− 972	− 430

9.

8,002	9,003	4,005	3,006	8,007	1,006
− 628	− 324	− 556	− 439	− 299	− 307

10.

4,003	6,002	3,008	9,004	5,005	8,003
− 2,564	− 3,574	− 2,569	− 2,916	− 4,936	− 1,227

EXAMPLE 1

$$\begin{array}{r} 6,000 \\ -2,786 \end{array}$$

$$\begin{array}{r} {}^{5}\cancel{6},{}^{9}\cancel{0}{}^{9}\cancel{0}{}^{10}0 \\ -2,786 \\ \hline 3,214 \\ \hline 6,000 \end{array}$$

EXAMPLE 2

$$\begin{array}{r} 60,704 \\ -59,857 \end{array}$$

$$\begin{array}{r} {}^{5}\cancel{6}{}^{9}\cancel{0},{}^{6}\cancel{7}{}^{10}\cancel{0}{}^{1}4 \\ -59,857 \\ \hline 847 \\ \hline 60,704 \end{array}$$

Subtract and check.

1.

5,000	4,000	8,000	1,000	6,000	7,000
− 436	− 228	− 927	− 533	− 407	− 316

2.

9,000	7,000	2,000	4,000	3,000	5,000
− 2,544	− 1,633	− 1,429	− 2,338	− 1,426	− 3,258

3.

2,040	3,004	5,000	6,200	3,060	4,002
− 1,537	− 1,252	− 2,430	− 2,576	− 1,485	− 3,268

4.

20,000	40,000	30,000	50,000	70,000
− 18,954	− 25,364	− 11,456	− 23,812	− 42,307

5.

40,007	80,006	70,003	10,004	50,002
− 23,564	− 51,288	− 29,436	− 9,543	− 23,875

Rewrite the problems on this page with the larger number on top and the smaller number directly below it. Put units under units, tens under tens, and so on.

1. $407 - 98 =$ $602 - 55 =$ $902 - 27 =$

2. $800 - 28 =$ $500 - 73 =$ $300 - 44 =$

3. $700 - 256 =$ $600 - 421 =$ $400 - 109 =$

4. $4,080 - 493 =$ $6,070 - 576 =$ $2,050 - 288 =$

5. $3,090 - 1,987 =$ $7,050 - 3,456 =$ $6,020 - 2,071 =$

6. $8,004 - 438 =$ $4,003 - 927 =$ $9,002 - 605 =$

7. $20,000 - 2,054 =$ $60,000 - 3,118 =$ $40,000 - 6,417 =$

Subtraction Shortcuts

When the subtrahend (the number you subtract from another number) ends in zero, you can subtract in your head. Any number minus zero is that number.

EXAMPLE 1   46 – 30 =

> **STEP 1**   Do not rewrite the problem. You know that 6 – 0 = 6.
>
> **STEP 2**   You also know that 40 – 30 = 10.
>
> **STEP 3**   The answer to 46 – 30 = 16.
>
> **STEP 4**   Check by adding 16 + 30 = 46.

 Subtract each problem in your head.

1. 72 – 40 =   97 – 60 =   123 – 80 =

2. 38 – 10 =   41 – 30 =   156 – 90 =

3. 65 – 20 =   84 – 50 =   139 – 70 =

Sometimes you can change a subtraction problem to a similar problem that is easier to subtract. Look at the next example carefully.

EXAMPLE 2   94 – 67 =

> **STEP 1**   Add 3 to both numbers to make   94 – 67 =
> the subtrahend end with zero.   + 3   + 3
>   ⎯⎯⎯⎯⎯⎯⎯⎯
> **STEP 2**   Subtract the new numbers.   97 – 70 = 27
>
> **STEP 3**   Check by subtracting 67 from 94.   94 – 67 = 27

Rewrite each problem as a similar problem with a subtrahend that ends in zero. Remember to add to *both* numbers in the problem. Then subtract the new numbers.

4. 83 – 26 =   104 – 88 =   52 – 19 =

5. 92 – 45 =   71 – 35 =   114 – 79 =

6. 90 – 63 =   85 – 56 =   64 – 28 =

7. 96 – 38 =   85 – 17 =   131 – 47 =

Rounding, Estimating, and Using a Calculator

Review the rules for rounding numbers on page 11 and the example on page 25. Use your skills in rounding numbers to estimate the answers to the problems on this page and the next.

Round each number to the nearest *hundred*. Then subtract the rounded numbers.

1. 42,356 − 9,485

2. 38,182 − 8,491

3. 87,421 − 9,530

4. 78,253 − 9,682

5. 34,818 − 5,847

6. 92,309 − 8,958

7. 69,762 − 4,974

8. 86,583 − 5,999

9. 47,757 − 6,858

10. 81,912 − 1,943

Round each number to the nearest *thousand*. Then subtract the rounded numbers.

11. 22,483 − 19,584

12. 62,537 − 56,849

13. 55,130 − 43,273

14. 25,614 − 16,726

15. 678,054 − 97,347

16. 327,583 − 58,946

17. 227,150 − 82,366

18. 732,122 − 51,475

19. 923,644 − 482,637

20. 482,637 − 297,148

21. 822,472 − 631,584

22. 673,612 − 297,641

Round each number to the nearest *ten thousand*. Then subtract the rounded numbers.

23. 208,625 − 120,432

24. 524,088 − 418,950

25. 809,307 − 326,175

26. 473,162 − 213,076

27. 398,416 − 106,432

28. 189,268 − 98,416

Using a Calculator A calculator is a convenient tool for getting an exact answer.

EXAMPLE   Use a calculator to find the answer to 2,415 − 398.

Press the following keys on a calculator: `2` `4` `1` `5` `−` `3` `9` `8` `=`

The calculator display should read   2017.

 For problems 29 to 34 round each number to the nearest *thousand* and subtract the rounded numbers. Then use a calculator to find the exact answers.

29. 74,925 − 23,936 =

30. 53,286 − 42,755 =

31. 42,755 − 22,697 =

32. 54,278 − 32,489 =

33. 45,718 − 21,279 =

34. 28,786 − 19,797 =

 For problems 35 to 38 round each number to the nearest *hundred thousand* and subtract the rounded numbers. Then use a calculator to find the exact answers.

35. 2,543,950 − 418,513 =

36. 6,473,032 − 1,214,877 =

37. 9,144,480 − 1,342,895 =

38. 2,738,259 − 2,589,767 =

Applying Your Subtraction Skills

On this page and the following four pages are problems that require you to apply your subtraction skills to practical problems. In each problem, pay close attention to the language that tells you to subtract. Watch for words such as **difference, balance, how many more, how much larger,** and **how much change.**

Line up money problems carefully. For example, to find how much change you would get from $10.00 for a $5.79 purchase, set the numbers up this way:

$$\begin{array}{r} \$10.00 \\ -\ \ 5.79 \\ \hline \$\ \ 4.21 \end{array}$$

Solve and write the correct label, such as $, miles, or pounds, next to each answer.

1. Carlos' schoolbooks came to a total of $38.45. How much change should he get from $50.00?

2. In Phoenix, Arizona, the lowest temperature on record is 16°. The highest temperature on record is 118°. What is the difference between the highest and lowest temperatures?

3. A pair of boots on sale were marked down from $65.00 to $49.50. How much can Imani save by buying the boots on sale?

4. In Midvale, the average annual income for a family with two wage earners is $48,231. For families with only one wage earner the average is $27,649. How much greater is the average income of a two-earner family than the average income for a one-earner family?

5. Unrestricted round-trip air fare between New York and Orlando is $702. Discount fare between the two cities is $192. What is the difference between the unrestricted fare and the discount fare?

6. Mr. and Mrs. Ford made a down payment of $13,425 on a house. The purchase price of the house was $89,500. The balance left after making the down payment is the amount of their mortgage. Find the amount of the mortgage.

7. A new housing project has 2,264 apartments. Before the project opened, 1,452 of the apartments were already rented. How many apartments were still not rented when the project opened?

8. Mrs. Garrett's groceries came to a total of $42.56. How much change should she get if she gives the cashier $50.00?

9. In his baseball career, Pete Rose made a total of 4,256 hits. Ty Cobb made 4,191 hits. Rose made how many more hits than Cobb?

10. The Mississippi River is 1,171 miles long. The Ohio River is 981 miles long. How much longer is the Mississippi River than the Ohio River?

Use the following information to answer questions 11 and 12.

The area of North America is 8,306,985 square miles. The area of South America is 7,940,960 square miles.

11. Round each number to the nearest thousand. Then use the rounded numbers to estimate how much larger North America is than South America.

12. Use a calculator to find exactly how much larger North America is than South America.

13. A mountain bike originally sold for $290.00. The bike was on sale for $198.80. The sale price is how much less than the original price?

14. In Orangeburg, Florida, the average annual income of a family with a male head of household is $21,162. The average income for a family with a female head of household is $11,589. How much more is the average income for a family with a male head than for a family with a female head?

15. The highest mountain in the world, Mount Everest, is 29,028 feet high. The second highest mountain, K2, is 28,250 feet high. Mount Everest is how much higher than K2?

16. The total weight of the members of the Tyson family was 922 pounds. Everyone in the family decided to diet. One month later their total weight was 878 pounds. How much weight did the family lose?

17. The Rosa family takes home $2,108.00 every month. They pay $495.70 each month for rent. How much do they have left for other expenses after paying rent?

Read the next problems carefully. Some of them require addition, and others require subtraction.

18. A wool shirt was on sale for $22.95. This was $8.05 less than the original price. What was the original price of the shirt?

Use the following information to answer questions 19 to 21.

The area of China is 3,696,100 square miles. The area of India is 1,269,340 square miles. The area of Iran is 632,457 square miles. The area of Pakistan is 339,697 square miles.

 19. Round each area to the nearest ten thousand square miles. Then use the rounded numbers to estimate the combined area of India, Iran, and Pakistan.

20. Is the combined area of India, Iran, and Pakistan more or less than the area of China?

 21. Use a calculator to find the difference between the area of China and the combined areas of India, Iran, and Pakistan.

22. Lois bought food for her party for $87.63. She bought beverages for the party for $36.88. She also bought paper plates, paper napkins, and plastic cups for $12.56. How much did these party items cost altogether?

23. Which of the following is the *smallest* amount of money that is enough to cover the total cost of the items for Lois' party?

   a. $120
   b. $130
   c. $140
   d. $150

24. An inn in Pennsylvania first opened for business in 1782. The inn stayed in business over the years. For how many years had the inn been operating by the year 2000?

25. One year the U.S. produced 9,335,227 automobiles. That year Germany produced 2,440,448 automobiles. To the nearest million, how many more automobiles did the U.S. produce than Germany?

   a. 5,000,000
   b. 6,000,000
   c. 7,000,000
   d. 8,000,000

26. Use a calculator to find exactly how many more automobiles the U.S. produced than Germany.

27. In 1990 the population of the U.S. was 248,143,000. According to some estimates, the population will be 296,511,000 in the year 2010. If the estimates are correct, which of the following is closest to the increase in population of the U.S. from 1990 to 2010?

   a. 500,000
   b. 5,000,000
   c. 50,000,000
   d. 500,000,000

28. One year a chain of music stores in Chicago sold 586,412 rock and roll CDs; 267,338 country music CDs; 75,056 jazz and blues CDs; 12,409 classical CDs; and 43,870 CDs of other categories. Did the total sale of CDs pass the 1,000,000 mark? (Hint: Round each number to the nearest ten thousand.)

29. Use a calculator to find the exact number of CDs that the chain sold.

Subtraction Review

This review covers the material you have studied so far in this book.
Read the signs (+ and −) carefully.

1. What is the value of the digit 8 in the number 1,839,567?

2. The population of Madison County is 249,538. What is the
   population rounded to the nearest ten thousand?

3. 873
   − 452

4. 968
   − 202

5. 637
   − 49

6. 536
   − 287

7. 4,850
   − 2,639

8. 800
   − 492

9. 7,002
   − 2,567

10. 20,030
    − 4,038

11. 300,000
    − 214,500

12. $2,784 + 1,956 =$

13. $61 + 86 + 49 + 52 =$

14. $9,238 − 479 =$

15. $406,000 − 209,513 =$

16. $60,080 − 4,236 =$

17. $70,005 − 38,028 =$

18. Round each number to the nearest *thousand* and subtract.

   $348,214 − 198,247$

19. Round each number to the nearest *ten thousand* and subtract.

   $894,328 − 59,069$

20. Round each number to the nearest *hundred thousand* and subtract.

2,456,320 − 1,804,317

21. In a recent year, there were 19,227 immigrants from the former Soviet Union to New York state. The same year there were 10,045 immigrants from the former Soviet Union to California. The number of immigrants to New York state was how many more than the number of immigrants to California?

22. Ethel sold her car for $4,988. This was $1,437 less than she paid for the car when she bought it a year ago. How much did she pay for the car?

23. The Millers bought new furniture for $430.00. If they made a down payment of $145.50, how much do they have left to pay?

Use the following information to answer questions 24 and 25.

One year France produced 3,050,929 cars and Italy produced 1,422,359 cars.

24. Round each number to the nearest hundred thousand. About how many more cars were produced in France than in Italy?

25. Together, how many cars did France and Italy produce that year?

Subtraction Review Chart

If you missed more than one problem on any group below, review the practice pages for those problems. Then redo the problems you got wrong before going on to the Multiplication Skills Inventory. If you had a passing score, redo any problem you missed and begin the Multiplication Skills Inventory on page 60.

PROBLEM NUMBERS	SKILL AREA	PRACTICE PAGES
1	place value	6–10
2	rounding whole numbers	11–12
3, 4	subtraction facts	35–38
5, 6, 7	subtracting and borrowing	39–43
8, 9, 10, 11	subtracting from zeros	45–48
12, 13	adding horizontally	23
14, 15, 16, 17	subtracting horizontally	44, 48
18, 19, 20	rounding and estimating	50–51
21, 23, 24	applying subtraction	52–56
22, 25	applying addition	27–30

MULTIPLICATION

Multiplication Skills Inventory

Do all the problems that you can. There is no time limit. Work carefully and check your answers, but do not use outside help.

1. 42
   × 3

2. 61
   × 5

3. 723
   × 3

4. 8,102
   × 4

5. 52
   × 23

6. 910
   × 62

7. 512 × 34 =

8. 6,122 • 43 =

9. 42
   × 8

10. (236)(9) =

11. 84
   × 37

12. 87
   × 46

13. 5,936(27) =

14. 6,254 • 83 =

15. (8,395)(516) =

16. 92,047 × 506 =

17. 38,520 • 780 =

18. 47 × 100 =

19. $1,000(526) =$ **20.** $(4,936)(279) =$ **21.** $638 \times 5,076 =$

22. Anna runs 4 miles per day 364 days a year. Round 364 to the nearest ten and estimate the number of miles Anna runs in a year.

23. Fumio makes $1,685 every month. How much does he make in one year?

24. A plane flies at an average speed of 413 miles per hour for 9 hours. How far does the plane fly in that time?

25. On a map of the United States, 1 inch represents 280 miles. How far apart are two cities that are 7 inches apart on the map?

Multiplication Skills Inventory Chart

If you missed more than one problem on any group below, work through the practice pages for that group. Then redo the problems you got wrong on the Multiplication Skills Inventory. If you had a passing score on all five groups of problems, redo any problem you missed and begin the Division Skills Inventory on page 87.

PROBLEM NUMBERS	SKILL AREA	PRACTICE PAGES
1, 2, 3, 4	multiplication facts	62–66
5, 6, 7, 8	multiplying larger numbers	67–70
9, 10, 11, 12, 13, 14 15, 16, 17, 18, 19, 20, 21	multiplying and carrying	71–75
22	rounding and estimating	76–78
23, 24, 25	applying multiplication	79–83

Basic Multiplication Facts

This page and the following two pages will help you learn the basic multiplication facts. These are more examples of **mental math.** The multiplication facts are basic building blocks for further study of mathematics. You must *memorize* these facts. Check your answers to this exercise. Practice the facts you got wrong. Then try these problems again until you can get all the answers quickly and accurately.

Parts of a Multiplication Problem

$$
\begin{array}{r}
82 \\
\times \ 3 \\
\hline
246
\end{array}
\quad
\begin{array}{l}
\longleftarrow \text{ multiplicand} \\
\longleftarrow \text{ multiplier} \\
\longleftarrow \text{ product}
\end{array}
\ \Big\}\ \text{These numbers are also called \textbf{factors.}}
$$

1. $8 \times 4 =$	$1 \times 11 =$	$7 \times 8 =$	$3 \times 7 =$	$0 \times 8 =$
2. $5 \times 12 =$	$9 \times 2 =$	$7 \times 12 =$	$8 \times 2 =$	$2 \times 11 =$
3. $4 \times 5 =$	$9 \times 6 =$	$5 \times 6 =$	$6 \times 8 =$	$4 \times 11 =$
4. $4 \times 3 =$	$11 \times 12 =$	$6 \times 0 =$	$7 \times 7 =$	$1 \times 12 =$
5. $2 \times 12 =$	$8 \times 8 =$	$3 \times 4 =$	$3 \times 6 =$	$5 \times 8 =$
6. $9 \times 5 =$	$10 \times 12 =$	$11 \times 11 =$	$8 \times 6 =$	$6 \times 5 =$
7. $10 \times 11 =$	$9 \times 9 =$	$9 \times 4 =$	$6 \times 12 =$	$3 \times 11 =$
8. $6 \times 7 =$	$4 \times 4 =$	$4 \times 8 =$	$12 \times 11 =$	$3 \times 9 =$
9. $7 \times 9 =$	$7 \times 1 =$	$9 \times 11 =$	$8 \times 12 =$	$5 \times 11 =$
10. $7 \times 6 =$	$1 \times 5 =$	$4 \times 6 =$	$8 \times 3 =$	$3 \times 12 =$
11. $8 \times 9 =$	$5 \times 5 =$	$6 \times 6 =$	$4 \times 12 =$	$8 \times 11 =$
12. $8 \times 7 =$	$7 \times 4 =$	$9 \times 8 =$	$7 \times 11 =$	$10 \times 10 =$

13. $6 \times 7 =$ $9 \times 9 =$ $9 \times 5 =$ $10 \times 11 =$ $8 \times 8 =$

14. $11 \times 11 =$ $3 \times 6 =$ $1 \times 12 =$ $5 \times 8 =$ $9 \times 4 =$

15. $2 \times 12 =$ $5 \times 0 =$ $10 \times 12 =$ $7 \times 7 =$ $4 \times 3 =$

16. $6 \times 5 =$ $11 \times 12 =$ $5 \times 6 =$ $3 \times 4 =$ $6 \times 8 =$

17. $2 \times 11 =$ $4 \times 5 =$ $9 \times 2 =$ $4 \times 11 =$ $9 \times 6 =$

18. $7 \times 12 =$ $3 \times 7 =$ $8 \times 2 =$ $0 \times 3 =$ $8 \times 4 =$

19. $1 \times 11 =$ $5 \times 12 =$ $7 \times 8 =$ $9 \times 7 =$ $7 \times 3 =$

20. $6 \times 9 =$ $12 \times 12 =$ $6 \times 11 =$ $5 \times 9 =$ $9 \times 8 =$

21. $7 \times 5 =$ $9 \times 3 =$ $10 \times 10 =$ $9 \times 12 =$ $7 \times 11 =$

22. $8 \times 11 =$ $8 \times 7 =$ $5 \times 5 =$ $7 \times 4 =$ $6 \times 6 =$

23. $8 \times 3 =$ $4 \times 12 =$ $7 \times 6 =$ $8 \times 9 =$ $3 \times 12 =$

24. $1 \times 5 =$ $9 \times 11 =$ $4 \times 6 =$ $8 \times 12 =$ $3 \times 9 =$

25. $5 \times 11 =$ $7 \times 9 =$ $4 \times 4 =$ $6 \times 1 =$ $4 \times 8 =$

26. $6 \times 12 =$ $12 \times 11 =$ $3 \times 11 =$ $8 \times 6 =$ $9 \times 8 =$

27. $6 \times 9 =$ $5 \times 9 =$ $7 \times 5 =$ $9 \times 12 =$ $7 \times 3 =$

28. $9 \times 3 =$ $12 \times 12 =$ $9 \times 7 =$ $4 \times 7 =$ $6 \times 11 =$

29.
12	9	5	7	3	12	9
× 9	×8	×5	×6	×9	× 6	×5

30.
12	4	11	3	11	10	12
× 1	×3	× 2	×7	× 6	×10	× 4

31.
4	7	6	11	12	3	9
×6	×1	×7	×11	×10	×4	×6

32.
12	7	11	6	1	7	11
× 5	×3	× 7	×6	×5	×9	× 3

33.
11	12	12	4	0	9	6
×10	× 2	×11	×5	×6	×3	×9

34.
11	8	11	4	8	8	9
× 8	×3	× 9	×4	×6	×8	×0

35.
5	9	8	12	5	8	12
×6	×2	×4	×12	×9	×7	× 3

36.
12	9	3	7	6	12	11
× 8	×9	×6	×7	×8	× 7	× 1

37.
9	7	8	11	7	4	12
×7	×5	×9	× 5	×4	×8	×11

38.
9	5	6	11	8	7	9
×4	×8	×5	× 4	×2	×8	×12

The Multiplication Table

It is very important to know the multiplication table. If you do not know it, take the time to memorize it. The time you spend memorizing the table now will be saved later on because you will be able to do long multiplication and division problems quickly.

This table does not include 0 because there is only one thing to remember when you multiply by 0: any number multiplied by 0 is 0.

	1	2	3	4	5	6	7	8	9	10	11	12
1	1	2	3	4	5	6	7	8	9	10	11	12
2	2	4	6	8	10	12	14	16	18	20	22	24
3	3	6	9	12	15	18	21	24	27	30	33	36
4	4	8	12	16	20	24	28	32	36	40	44	48
5	5	10	15	20	25	30	35	40	45	50	55	60
6	6	12	18	24	30	36	42	48	54	60	66	72
7	7	14	21	28	35	42	49	56	63	70	77	84
8	8	16	24	32	40	48	56	64	72	80	88	96
9	9	18	27	36	45	54	63	72	81	90	99	108
10	10	20	30	40	50	60	70	80	90	100	110	120
11	11	22	33	44	55	66	77	88	99	110	121	132
12	12	24	36	48	60	72	84	96	108	120	132	144

Reading the Multiplication Table

Look at the row of numbers running across the top of the table, and then look at the column of numbers running along the left side. You can multiply any number in the top row by any number in the column on the left and find the answer inside the table. For example, to find out how much 8 × 7 is, locate the number 8 in the top row and the number 7 in the column on the left. Run your finger down the 8 column until you have reached the row marked 7. The answer is 56. You also could have found the answer by starting at the 8 on the left and moving your finger across the table until you reached the vertical column of figures marked 7 at the top. The multiplication table can be read both horizontally and vertically.

Notice also that the column of numbers under the 7 and the row of numbers next to the 7 increase by seven each time (7 + 7 = 14; 14 + 7 = 21; 21 + 7 = 28; and so on).

Multiplying by One-Digit Numbers

Multiply each digit in the top number by the bottom number. The answer should be written from right to left, starting with the product of the ones column, then the tens column, then the hundreds column, and so on.

EXAMPLE 1
$$\begin{array}{r} 243 \\ \times\ 2 \\ \hline 486 \end{array}$$

STEP 1 $2 \times 3 = 6$
STEP 2 $2 \times 4 = 8$
STEP 3 $2 \times 2 = 4$

Checking a multiplication problem is sometimes more difficult than actually working the problem. One method is to go over your steps to try to find any errors you might have made. Another more reliable method is to divide the answer you got by the bottom number in the problem. If you get the top number in the original problem as the answer to the division problem, your multiplication is correct. If you are not sure of your division at this point, don't worry about using this method of checking a multiplication problem. You will begin practicing division on page 90.

EXAMPLE 2
$$\begin{array}{r} 243 \\ \times\ 2 \\ \hline 486 \end{array}$$

CHECK
$$2\overline{)486}^{243}$$

Multiply and check.

1.
$$\begin{array}{r} 71 \\ \times\ 9 \\ \hline \end{array} \qquad \begin{array}{r} 81 \\ \times\ 7 \\ \hline \end{array} \qquad \begin{array}{r} 62 \\ \times\ 4 \\ \hline \end{array} \qquad \begin{array}{r} 43 \\ \times\ 3 \\ \hline \end{array} \qquad \begin{array}{r} 91 \\ \times\ 8 \\ \hline \end{array} \qquad \begin{array}{r} 90 \\ \times\ 3 \\ \hline \end{array} \qquad \begin{array}{r} 20 \\ \times\ 8 \\ \hline \end{array} \qquad \begin{array}{r} 70 \\ \times\ 7 \\ \hline \end{array}$$

2.
$$\begin{array}{r} 312 \\ \times\ 4 \\ \hline \end{array} \qquad \begin{array}{r} 821 \\ \times\ 3 \\ \hline \end{array} \qquad \begin{array}{r} 611 \\ \times\ 9 \\ \hline \end{array} \qquad \begin{array}{r} 401 \\ \times\ 6 \\ \hline \end{array} \qquad \begin{array}{r} 502 \\ \times\ 4 \\ \hline \end{array} \qquad \begin{array}{r} 601 \\ \times\ 9 \\ \hline \end{array} \qquad \begin{array}{r} 801 \\ \times\ 7 \\ \hline \end{array}$$

3.
$$\begin{array}{r} 610 \\ \times\ 7 \\ \hline \end{array} \qquad \begin{array}{r} 420 \\ \times\ 3 \\ \hline \end{array} \qquad \begin{array}{r} 510 \\ \times\ 8 \\ \hline \end{array} \qquad \begin{array}{r} 110 \\ \times\ 9 \\ \hline \end{array} \qquad \begin{array}{r} 700 \\ \times\ 5 \\ \hline \end{array} \qquad \begin{array}{r} 900 \\ \times\ 7 \\ \hline \end{array} \qquad \begin{array}{r} 200 \\ \times\ 6 \\ \hline \end{array}$$

4.
$$\begin{array}{r} 4,201 \\ \times\ 4 \\ \hline \end{array} \qquad \begin{array}{r} 3,102 \\ \times\ 3 \\ \hline \end{array} \qquad \begin{array}{r} 5,021 \\ \times\ 2 \\ \hline \end{array} \qquad \begin{array}{r} 6,011 \\ \times\ 8 \\ \hline \end{array} \qquad \begin{array}{r} 7,101 \\ \times\ 6 \\ \hline \end{array} \qquad \begin{array}{r} 8,011 \\ \times\ 7 \\ \hline \end{array}$$

Multiplying by Larger Numbers

When you multiply by a 2-digit number, be sure to begin your answer (the first partial product) directly under the ones columns of the numbers being multiplied. Begin the second partial product directly under the tens columns of the numbers being multiplied and the first partial product. Always multiply from right to left.

EXAMPLE

$$
\begin{array}{r}
51 \\
\times 32 \\
\hline
102 \\
153 \\
\hline
1{,}632
\end{array}
$$

partial products $\{$

STEP 1 $2 \times 1 = 2$
STEP 2 $2 \times 5 = 10$
STEP 3 $3 \times 1 = 3$
STEP 4 $3 \times 5 = 15$
STEP 5 Add the partial products to get the final product.

Multiply and check.

1.
$$
\begin{array}{r} 62 \\ \times 23 \\ \hline \end{array}
\qquad
\begin{array}{r} 23 \\ \times 31 \\ \hline \end{array}
\qquad
\begin{array}{r} 31 \\ \times 57 \\ \hline \end{array}
\qquad
\begin{array}{r} 92 \\ \times 43 \\ \hline \end{array}
\qquad
\begin{array}{r} 24 \\ \times 22 \\ \hline \end{array}
\qquad
\begin{array}{r} 52 \\ \times 34 \\ \hline \end{array}
\qquad
\begin{array}{r} 81 \\ \times 75 \\ \hline \end{array}
$$

2.
$$
\begin{array}{r} 42 \\ \times 44 \\ \hline \end{array}
\qquad
\begin{array}{r} 71 \\ \times 89 \\ \hline \end{array}
\qquad
\begin{array}{r} 51 \\ \times 63 \\ \hline \end{array}
\qquad
\begin{array}{r} 43 \\ \times 21 \\ \hline \end{array}
\qquad
\begin{array}{r} 72 \\ \times 41 \\ \hline \end{array}
\qquad
\begin{array}{r} 91 \\ \times 66 \\ \hline \end{array}
\qquad
\begin{array}{r} 61 \\ \times 58 \\ \hline \end{array}
$$

3.
$$
\begin{array}{r} 50 \\ \times 73 \\ \hline \end{array}
\qquad
\begin{array}{r} 30 \\ \times 29 \\ \hline \end{array}
\qquad
\begin{array}{r} 90 \\ \times 64 \\ \hline \end{array}
\qquad
\begin{array}{r} 70 \\ \times 52 \\ \hline \end{array}
\qquad
\begin{array}{r} 20 \\ \times 18 \\ \hline \end{array}
\qquad
\begin{array}{r} 80 \\ \times 47 \\ \hline \end{array}
\qquad
\begin{array}{r} 40 \\ \times 36 \\ \hline \end{array}
$$

4.
$$
\begin{array}{r} 901 \\ \times 56 \\ \hline \end{array}
\qquad
\begin{array}{r} 510 \\ \times 68 \\ \hline \end{array}
\qquad
\begin{array}{r} 5{,}111 \\ \times 89 \\ \hline \end{array}
\qquad
\begin{array}{r} 9{,}121 \\ \times 34 \\ \hline \end{array}
\qquad
\begin{array}{r} 7{,}122 \\ \times 43 \\ \hline \end{array}
\qquad
\begin{array}{r} 91{,}001 \\ \times 78 \\ \hline \end{array}
$$

When multiplying by zero, write the answer 0 directly under the 0 in the problem. Then multiply by the next digit in the bottom number and continue to the left. The first problem is done for you as an example.

....

5.
$$
\begin{array}{r} 82 \\ \times\ 40 \\ \hline 3{,}280 \end{array}
\qquad
\begin{array}{r} 71 \\ \times 80 \\ \hline \end{array}
\qquad
\begin{array}{r} 52 \\ \times 30 \\ \hline \end{array}
\qquad
\begin{array}{r} 91 \\ \times 50 \\ \hline \end{array}
\qquad
\begin{array}{r} 62 \\ \times 40 \\ \hline \end{array}
\qquad
\begin{array}{r} 43 \\ \times 30 \\ \hline \end{array}
\qquad
\begin{array}{r} 31 \\ \times 90 \\ \hline \end{array}
$$

6.
$$
\begin{array}{r} 641 \\ \times\ 20 \\ \hline \end{array}
\qquad
\begin{array}{r} 312 \\ \times\ 40 \\ \hline \end{array}
\qquad
\begin{array}{r} 511 \\ \times\ 70 \\ \hline \end{array}
\qquad
\begin{array}{r} 713 \\ \times\ 20 \\ \hline \end{array}
\qquad
\begin{array}{r} 912 \\ \times\ 30 \\ \hline \end{array}
\qquad
\begin{array}{r} 411 \\ \times\ 80 \\ \hline \end{array}
$$

....

Notice where the product of the hundreds column begins in problems with a 3-digit multiplier. Remember each partial product starts under and one column to the left of the previous partial product.

EXAMPLE
$$
\begin{array}{r}
\leftarrow \text{hundreds column} \\
10{,}321 \\
\times\ \ \ \ 113 \\
\hline
30\ 963 \\
103\ 21 \\
1\ 032\ 1 \\
\hline
1{,}166{,}273
\end{array}
\left.\right\} \text{partial products}
$$

....

7.
$$
\begin{array}{r} 31{,}021 \\ \times\ \ \ 213 \\ \hline \end{array}
\qquad
\begin{array}{r} 80{,}011 \\ \times\ \ \ 497 \\ \hline \end{array}
\qquad
\begin{array}{r} 60{,}112 \\ \times\ \ \ 314 \\ \hline \end{array}
\qquad
\begin{array}{r} 10{,}220 \\ \times\ \ \ 123 \\ \hline \end{array}
\qquad
\begin{array}{r} 71{,}011 \\ \times\ \ \ 856 \\ \hline \end{array}
$$

8.
$$
\begin{array}{r} 51{,}423 \\ \times\ \ \ 112 \\ \hline \end{array}
\qquad
\begin{array}{r} 40{,}203 \\ \times\ \ \ 232 \\ \hline \end{array}
\qquad
\begin{array}{r} 23{,}121 \\ \times\ \ \ 321 \\ \hline \end{array}
\qquad
\begin{array}{r} 12{,}332 \\ \times\ \ \ 302 \\ \hline \end{array}
\qquad
\begin{array}{r} 31{,}212 \\ \times\ \ \ 443 \\ \hline \end{array}
$$

Multiplying Numbers Written Horizontally

To multiply numbers written horizontally, rewrite the problem vertically with the shorter number on the bottom. Whether you put the shorter number on the top or on the bottom, you will get the same answer. However, by putting the shorter number on the bottom, you will have fewer partial products to add.

EXAMPLE $312 \times 24 =$

$$
\begin{array}{r}
312 \\
\times\ 24 \\
\hline
1\,248 \\
6\,24 \\
\hline
7{,}488
\end{array}
\left.\right\}
\qquad
\begin{array}{r}
24 \\
\times 312 \\
\hline
48 \\
24 \\
7\,2 \\
\hline
7{,}488
\end{array}
\left.\right\}
\text{partial products}
$$

Multiply and check.

1. $41 \times 3 =$ $81 \times 5 =$ $64 \times 2 =$ $53 \times 3 =$

2. $741 \times 2 =$ $323 \times 3 =$ $611 \times 5 =$ $421 \times 4 =$

3. $331 \times 23 =$ $622 \times 41 =$ $812 \times 34 =$ $913 \times 22 =$

4. $703 \times 33 =$ $97 \times 801 =$ $601 \times 38 =$ $76 \times 610 =$

5. $212 \times 33 =$ $43 \times 31 =$ $73 \times 22 =$ $93 \times 133 =$

So far in this book you have seen the $\times$ sign to indicate multiplication. There are three other common ways to indicate multiplication. Each of the following means "five times six is equal to thirty."

EXAMPLE 1 $5 \times 6 = 30$

EXAMPLE 2 $5 \cdot 6 = 30$ The raised dot means to multiply.

EXAMPLE 3 $5(6) = 30$ A number next to another number in parentheses means to multiply.

EXAMPLE 4 $(5)(6) = 30$ Numbers in parentheses with no sign between them mean to multiply.

Notice that there is not a sign, such as + or –, in $5(6)$ or $(5)(6)$. When you study algebra, you will learn that $5 + (6)$ means to add and that $(5) + (6)$ also means to add.

....

6. $941 \cdot 20 =$ $\qquad$ $713 \cdot 30 =$ $\qquad$ $811 \cdot 80 =$ $\qquad$ $912 \cdot 40 =$

7. $21(540) =$ $\qquad$ $34(620) =$ $\qquad$ $75(810) =$ $\qquad$ $32(930) =$

8. $(33)(9,131) =$ $\qquad$ $(3,421)(22) =$ $\qquad$ $(21)(6,124) =$ $\qquad$ $(7,123)(23) =$

9. $51,211 \cdot 32 =$ $\qquad$ $24(70,212) =$ $\qquad$ $(61,201)(31) =$

10. $40(2,021) =$ $\qquad$ $31 \cdot 81,202 =$ $\qquad$ $(51)(61,001) =$

11. $(60)(80,010) =$ $\qquad$ $21(40,321) =$ $\qquad$ $12 \cdot 43,223 =$

Multiplying and Carrying

Carrying in multiplication is very much like carrying in addition. Be sure to multiply *first* and *then* add the number being carried. Carrying is sometimes called **regrouping** or **renaming**.

EXAMPLE

$$\begin{array}{r} 49 \\ \times\ 63 \\ \hline 147 \\ 2\,94 \\ \hline 3{,}087 \end{array}$$

STEP 1 $3 \times 9 = 27$ Write the 7. Carry the 2.
STEP 2 $3 \times 4 = 12$ $12 + 2 = 14$ Write the 14.
STEP 3 $6 \times 9 = 54$ Write the 4. Carry the 5.
STEP 4 $6 \times 4 = 24$ $24 + 5 = 29$ Write the 29.
STEP 5 Add the partial products.

Multiply and check.

1.
$\begin{array}{r}17\\\times\ 4\\\hline\end{array}$
$\begin{array}{r}64\\\times\ 5\\\hline\end{array}$
$\begin{array}{r}95\\\times\ 7\\\hline\end{array}$
$\begin{array}{r}76\\\times\ 2\\\hline\end{array}$
$\begin{array}{r}83\\\times\ 9\\\hline\end{array}$
$\begin{array}{r}43\\\times\ 6\\\hline\end{array}$
$\begin{array}{r}66\\\times\ 3\\\hline\end{array}$
$\begin{array}{r}52\\\times\ 8\\\hline\end{array}$

2.
$\begin{array}{r}34\\\times\ 5\\\hline\end{array}$
$\begin{array}{r}85\\\times\ 2\\\hline\end{array}$
$\begin{array}{r}43\\\times\ 8\\\hline\end{array}$
$\begin{array}{r}18\\\times\ 7\\\hline\end{array}$
$\begin{array}{r}27\\\times\ 3\\\hline\end{array}$
$\begin{array}{r}93\\\times\ 4\\\hline\end{array}$
$\begin{array}{r}54\\\times\ 6\\\hline\end{array}$
$\begin{array}{r}48\\\times\ 9\\\hline\end{array}$

3.
$\begin{array}{r}69\\\times\ 2\\\hline\end{array}$
$\begin{array}{r}36\\\times\ 9\\\hline\end{array}$
$\begin{array}{r}24\\\times\ 6\\\hline\end{array}$
$\begin{array}{r}85\\\times\ 3\\\hline\end{array}$
$\begin{array}{r}26\\\times\ 5\\\hline\end{array}$
$\begin{array}{r}48\\\times\ 7\\\hline\end{array}$
$\begin{array}{r}74\\\times\ 4\\\hline\end{array}$
$\begin{array}{r}53\\\times\ 8\\\hline\end{array}$

4.
$\begin{array}{r}53\\\times\ 5\\\hline\end{array}$
$\begin{array}{r}22\\\times\ 6\\\hline\end{array}$
$\begin{array}{r}84\\\times\ 3\\\hline\end{array}$
$\begin{array}{r}66\\\times\ 2\\\hline\end{array}$
$\begin{array}{r}75\\\times\ 9\\\hline\end{array}$
$\begin{array}{r}93\\\times\ 7\\\hline\end{array}$
$\begin{array}{r}86\\\times\ 4\\\hline\end{array}$
$\begin{array}{r}47\\\times\ 8\\\hline\end{array}$

5.
$\begin{array}{r}86\\\times\ 6\\\hline\end{array}$
$\begin{array}{r}64\\\times\ 4\\\hline\end{array}$
$\begin{array}{r}77\\\times\ 3\\\hline\end{array}$
$\begin{array}{r}29\\\times\ 8\\\hline\end{array}$
$\begin{array}{r}45\\\times\ 7\\\hline\end{array}$
$\begin{array}{r}33\\\times\ 5\\\hline\end{array}$
$\begin{array}{r}97\\\times\ 2\\\hline\end{array}$
$\begin{array}{r}52\\\times\ 9\\\hline\end{array}$

6.
$\begin{array}{r}69\\\times\ 6\\\hline\end{array}$
$\begin{array}{r}26\\\times\ 8\\\hline\end{array}$
$\begin{array}{r}84\\\times\ 3\\\hline\end{array}$
$\begin{array}{r}55\\\times\ 5\\\hline\end{array}$
$\begin{array}{r}73\\\times\ 7\\\hline\end{array}$
$\begin{array}{r}46\\\times\ 9\\\hline\end{array}$
$\begin{array}{r}74\\\times\ 4\\\hline\end{array}$
$\begin{array}{r}18\\\times\ 2\\\hline\end{array}$

7.
97	43	76	66	38	57	13	28
× 3	× 9	× 4	× 3	× 8	× 6	× 7	× 5

8.
37	54	46	73	93	62	27	85
× 5	× 8	× 3	× 9	× 4	× 7	× 2	× 6

9.
76	38	26	47	19	53	64	93
× 6	× 2	× 8	× 5	× 4	× 9	× 3	× 7

10.
42	66	58	73	85	14	26	39
× 7	× 5	× 9	× 8	× 3	× 4	× 6	× 2

Remember: Any number multiplied by 0 is 0, and you must always multiply before you carry. If you remember this, you won't be tricked by the next row of problems on this page.

EXAMPLE   403
   × 9
   3,627

STEP 1 $9 \times 3 = 27$ Write the 7.   Carry the 2.
STEP 2 $9 \times 0 = 0$ $0 + 2 = 2$ Write the 2.
STEP 3 $9 \times 4 = 36$

11.
509	407	803	602	708	109	406
× 3	× 6	× 5	× 9	× 8	× 2	× 7

12.
237	198	724	193	825	773	294
× 5	× 4	× 7	× 6	× 8	× 4	× 7

13.
862	754	817	563	359	268	738
× 9	× 3	× 6	× 5	× 4	× 5	× 9

14. $(36)(27) =$ $(82)(58) =$ $(65)(43) =$ $(94)(29) =$

15. $87 \cdot 63 =$ $34 \cdot 74 =$ $68 \cdot 53 =$ $56 \cdot 48 =$

16. $26(59) =$ $55(62) =$ $86(37) =$ $39(42) =$

17. $56 \times 83 =$ $72 \times 76 =$ $84 \times 38 =$ $93 \times 66 =$

18. $(27)(73) =$ $(64)(29) =$ $(75)(66) =$ $(93)(88) =$

Remember to take the shortcut. Write the number with fewer digits on the bottom.

19. $43(255) =$ $56(367) =$ $623(79) =$ $784(38) =$

20. $68 \cdot 895 =$ $96 \cdot 273 =$ $408 \cdot 53 =$ $907 \cdot 27 =$

21. $(46)(305) =$ $(604)(89) =$ $(73)(805) =$ $(24)(706) =$

22. $9,173 \times 63 =$ $27 \times 4,218 =$ $7,516 \times 54 =$ $73 \times 4,329 =$

23. $82(1,256) =$ $56(3,709) =$ $6,048(81) =$ $8,690(29) =$

Multiplying by 10, 100, and 1,000

To multiply a number by 10, add a 0 to the right of the number.

EXAMPLE 1 $25 \times 10 = 250$ or

$$
\begin{array}{r}
25 \\
\times\, 10 \\
\hline
250
\end{array}
$$

To multiply a number by 100, add two 0's to the right of the number.

EXAMPLE 2 $36 \times 100 = 3{,}600$ or

$$
\begin{array}{r}
36 \\
\times\, 100 \\
\hline
3{,}600
\end{array}
$$

To multiply a number by 1,000, add three 0's to the right of the number.

EXAMPLE 3 $721 \times 1{,}000 = 721{,}000$ or

$$
\begin{array}{r}
721 \\
\times\, 1{,}000 \\
\hline
721{,}000
\end{array}
$$

1.

47	53	92	36	71	28	65
× 10	× 10	× 10	× 10	× 10	× 10	× 10

2.

368	866	761	946	479	261
× 10	× 10	× 10	× 10	× 10	× 10

3.

9,483	2,356	3,079	4,308	5,570	6,090
× 100	× 100	× 100	× 100	× 100	× 100

4.

100	100	100	100	100	100
× 74	× 8	× 256	× 31	× 209	× 68

5.

1,000	1,000	1,000	1,000	1,000	1,000
× 73	× 421	× 16	× 208	× 450	× 623

6. $74 \times 10 =$ $31 \times 10 =$ $25 \times 10 =$ $56 \times 10 =$

7. $10 \times 98 =$ $10 \times 46 =$ $10 \times 50 =$ $10 \times 37 =$

8. $662 \times 10 =$ $296 \times 10 =$ $802 \times 10 =$ $429 \times 10 =$

9. $506 \times 10 =$ $380 \times 10 =$ $409 \times 10 =$ $110 \times 10 =$

10. $10 \times 128 =$ $10 \times 839 =$ $10 \times 756 =$ $10 \times 217 =$

11. $6{,}578 \times 10 =$ $7{,}308 \times 10 =$ $8{,}815 \times 10 =$ $7{,}049 \times 10 =$

12. $26 \times 100 =$ $43 \times 100 =$ $94 \times 100 =$ $37 \times 100 =$

13. $100 \times 95 =$ $100 \times 30 =$ $100 \times 63 =$ $100 \times 74 =$

14. $100 \times 81 =$ $100 \times 57 =$ $100 \times 68 =$ $100 \times 25 =$

15. $957 \times 100 =$ $214 \times 100 =$ $693 \times 100 =$ $898 \times 100 =$

16. $581 \times 100 =$ $142 \times 100 =$ $813 \times 100 =$ $420 \times 100 =$

17. $100 \times 225 =$ $100 \times 730 =$ $100 \times 207 =$ $100 \times 396 =$

18. $1{,}000 \times 65 =$ $1{,}000 \times 88 =$ $1{,}000 \times 62 =$ $1{,}000 \times 91 =$

19. $32 \times 1{,}000 =$ $603 \times 1{,}000 =$ $540 \times 1{,}000 =$ $931 \times 1{,}000 =$

20. $1{,}000 \times 46 =$ $1{,}000 \times 261 =$ $1{,}000 \times 380 =$ $1{,}000 \times 715 =$

21. $425 \times 1{,}000 =$ $1{,}000 \times 689 =$ $237 \times 1{,}000 =$ $1{,}000 \times 499 =$

Rounding, Estimating, and Using a Calculator

In the last exercise, you learned to multiply numbers by 10, 100, and 1,000 in your head. Zeros make many multiplication problems easier.

EXAMPLE 1 $6 \times 40 =$

            $6 \times 40 = 240$

STEP 1 Forget about the zero in 40, and multiply $6 \times 4 = 24$.

STEP 2 Bring along the 0 from 40.

EXAMPLE 2 $80(700) =$

            $80(700) = 56,000$

STEP 1 Forget about the zeros in 80 and 700, and multiply $8 \times 7 = 56$.

STEP 2 Bring along the zeros from 80 and 700.

Multiply each problem in your head.

1. $40 \times 8 =$ $900 \times 60 =$ $(1200)(30) =$

2. $9(70) =$ $5(700) =$ $9 \cdot 6,000 =$

3. $(60)(5) =$ $(20)(800) =$ $(2)(13,000) =$

4. $70 \cdot 20 =$ $5 \times 400 =$ $7,000 \times 80 =$

To estimate an answer to a multiplication problem, try rounding the larger number to the *left-most* place. Sometimes this is called **front-end rounding.**

For example, if the larger number in a problem is 427, round 427 to the nearest hundred. If the larger number is 2,846, round 2,846 to the nearest thousand.

EXAMPLE 1 Estimate the answer to 9×427.

      $9 \times 427 \approx 9 \times 400$

STEP 1 Round 427 to the nearest hundred. Remember the symbol $\approx$ means "is approximately equal to."

      $9 \times 400 = 3,600$

STEP 2 Multiply 9×400 in your head.

EXAMPLE 2 Estimate the answer to $5 \times 2,846$.

      $5 \times 2,846 \approx 5 \times 3,000$

STEP 1 Round 2,846 to the nearest thousand.

      $5 \times 3,000 = 15,000$

STEP 2 Multiply $5 \times 3,000$ in your head.

? Round the larger number in each problem to the nearest *hundred* and multiply.

5. $4 \times 782 \approx$ $7(284) \approx$ $3(447) \approx$

6. $(912)(3) \approx$ $609 \times 2 \approx$ $8 \cdot 931 \approx$

7. $6 \cdot 472 \approx$ $(5)(872) \approx$ $189 \times 4 \approx$

 Round the larger number in each problem to the nearest *thousand* and multiply.

8. $2(4,281) \approx$ $(7,516)(4) \approx$ $4(5,693) \approx$

9. $7 \times 2,963 \approx$ $5 \cdot 3,772 \approx$ $8 \cdot 1,204 \approx$

10. $3 \cdot 6,059 \approx$ $9,461 \times 6 \approx$ $(2)(18,366) \approx$

To get a quick estimate for an answer, you can use front-end rounding for both numbers.

In the next problems, round the left-most digit in *both* numbers and multiply the rounded numbers in your head. The first problem is started for you.

11. $28 \cdot 73 \approx 30 \cdot 70 =$ $57(243) \approx$ $(79)(4,123) \approx$

12. $12 \times 294 \approx$ $88 \times 32 \approx$ $231 \times 659 \approx$

13. $4,809 \cdot 71 \approx$ $726 \cdot 16 \approx$ $428 \cdot 973 \approx$

 Using a Calculator   A calculator is a convenient tool for getting an exact answer.

EXAMPLE   Use a calculator to find the product of 7×286.

Press the following keys on a calculator: [7] [×] [2] [8] [6] [=]

The calculator display should read [2002.]

In the following problems, use rounding to select the correct answer from the choices. Then use a calculator to check each answer.

14. $94 \times 81 =$

 a. 724
 b. 7,614
 c. 14,084
 d. 17,234

15. $4,156 \times 18 =$

 a. 41,208
 b. 56,608
 c. 74,808
 d. 83,308

16. $56(73) =$

 a. 4,088
 b. 8,208
 c. 14,084
 d. 14,518

17. $(39)(6,234) =$

 a. 18,646
 b. 24,316
 c. 186,246
 d. 243,126

18. $36 \cdot 425 =$

 a. 8,700
 b. 9,600
 c. 15,300
 d. 21,900

19. $26 \times 5,085 =$

 a. 132,210
 b. 112,910
 c. 90,670
 d. 88,560

20. $(78)(112) =$

 a. 3,226
 b. 4,216
 c. 6,106
 d. 8,736

21. $5,736 \cdot 63 =$

 a. 442,828
 b. 361,368
 c. 243,118
 d. 186,208

22. $963 \times 48 =$

 a. 28,804
 b. 37,514
 c. 39,084
 d. 46,224

23. $(472)(681) =$

 a. 152,642
 b. 181,822
 c. 243,512
 d. 321,432

24. $21(734) =$

 a. 9,364
 b. 15,414
 c. 18,704
 d. 20,064

25. $798 \cdot 657 =$

 a. 642,826
 b. 524,286
 c. 414,346
 d. 382,156

Applying Your Multiplication Skills

On this page and the following four pages are practical problems that require you to apply your multiplication skills. In each problem, pay close attention to the language that tells you to multiply. In most cases, you will be given information about one thing and you will be asked to apply it to several things.

EXAMPLE   If David makes $9.75 an hour, how much does he make in 8 hours?

$$\begin{array}{r} \$9.75 \\ \times \quad 8 \\ \hline \$78.00 \end{array}$$

Solve and write the correct label such as $, miles, or gallons next to each answer.

1. Maria drives her car at an average speed of 53 miles per hour. How far can she drive in 4 hours?

2. How much did Pat pay for 6 containers of orange juice if each container cost $1.99?

3. Vera can input 54 words per minute on her computer. How many words can she input in 12 minutes?

4. Mr. Torres worked 7 hours overtime last week. If he made $14.65 for each hour of overtime, how much was he paid for his overtime work?

5. A case of soda contains 24 cans. How many cans are there in 10 cases?

6. Joanne's station wagon gets 14 miles for every gallon of gasoline. If her tank holds 15 gallons, how far can she drive on a full tank of gasoline?

7. Clara gets $32 every evening that she works as a baby-sitter. How much does she make in 6 evenings?

Use the following information to answer questions 8 to 10.

Marvin owns a grocery store. He pays his supplier 59¢ for a can of corn. Marvin charges his customers 68¢ for a can of corn.

8. How much does Marvin pay his supplier for 500 cans of corn?

9. How much do Marvin's customers pay altogether for 500 cans of corn?

10. The difference between the price Marvin charges his customers and the price he pays his supplier is called the markup. What is the total markup on the 500 cans of corn?

11. A pilot flies his airplane at an average speed of 378 miles per hour. How far can he fly in 7 hours?

12. On a regional map 1 inch represents 65 miles. How many miles apart are two cities that are 5 inches apart on the map?

13. Floria bought three shirts for $26.79 each. Find the total cost for the three shirts.

14. Sound travels at 1,129 feet per second. How far away was a thunderbolt that took 8 seconds to reach the ear of a listener?

15. The Uptown Theatre has 43 rows of seats. Each row contains 57 seats. How many seats are there in the theatre?

16. Jeff saves $25 a week. How much can he save in one year?
    (1 year = 52 weeks)

Use the following information to answer questions 17 to 19.

Each building in the Roosevelt Housing Project has 16 floors of apartments. Each residential floor has 12 apartments.

17. How many apartments are there in each building in the project?

18. The entire Roosevelt Housing Project has 23 buildings. How many apartments are there in the project?

19. The average family in the project has 4 members. Which of the following is closest to the total number of residents in the Roosevelt Housing Project?

    a. 180 residents
    b. 1,800 residents
    c. 18,000 residents
    d. 180,000 residents

20. If a train travels at an average speed of 47 miles per hour, how far can it go in 12 hours?

21. In order to buy her furniture, Mrs. Donaldson agrees to pay $87 a month for 24 months. How much will the furniture cost her?

Use the following information to answer questions 22 to 24.

A clothing store owner pays $54.60 for a jacket. He charges his customers $69.95 for the jacket.

22. How much does the store owner pay for 20 jackets?

23. How much does the store owner get from his customers if he sells all 20 jackets?

24. What is the total markup on the 20 jackets?

25. Each classroom in Central School holds 33 students. There are 19 classrooms in the school. How many students can Central School accommodate?

26. There are 5,280 feet in one mile. How many feet are there in 23 miles?

27. To pay off their mortgage, the Smiths agree to pay $450 a month. How much will they pay toward their mortgage every year?

28. The Smiths, in the last problem, plan to continue to make monthly payments on their mortgage for 20 years. Find the entire amount that they will pay.

29. Mr. Ford runs a convenience store. He makes a profit of $0.07 on every 16-ounce can of juice. If he sold 583 cans of juice in a week, how much profit did he make?

30. It took 6 carpenters 160 hours each to build an addition to a community center. Find the total number of hours all the carpenters worked.

31. The average pay for each carpenter in the last problem was $15 an hour. What was the total cost to the community for the carpenters' work?

32. Light travels at a speed of 186,000 miles per second. About how far can light travel in 15 seconds?

    a. 30,000,000 miles
    b. 3,000,000 miles
    c. 300,000 miles
    d. 30,000 miles

Multiplication Review

This review covers the material you have studied so far in this book. Read the signs carefully to decide whether to add, subtract, or multiply.

 1. Jake made $9,482 at his part-time job last year. What was his income to the nearest hundred dollars?

2. What is the value of the 9 in $9,482?

3. $14 + 37 + 6 =$

4. $98,208 + 7,065 =$

5. $23,584 - 12,270 =$

6. $340,600 - 91,837 =$

7. $\begin{array}{r} 62 \\ \times\ 4 \\ \hline \end{array}$

8. $\begin{array}{r} 421 \\ \times\ \ 3 \\ \hline \end{array}$

9. $\begin{array}{r} 610 \\ \times\ 58 \\ \hline \end{array}$

10. $411 \times 36 =$

11. $(73)(6) =$

12. $8(457) =$

13. $85 \cdot 49 =$

14. $736 \cdot 53 =$

15. $83,056 \times 47 =$

16. $53 \times 1,000 =$

17. $100(423) =$

18. $7,240 \cdot 10 =$

19. Round both numbers to the nearest *hundred* and multiply.

(428)(782)

20. Round the smaller number to the nearest *hundred* and the larger number to the nearest *thousand*. Then multiply.

$469 \times 6,738$

21. Round the smaller number to the nearest *hundred* and the larger number to the nearest *ten thousand*. Then multiply.

(860)(49,630)

22. Mr. and Mrs. Burns bought a used car. They agreed to pay $120 every month for 28 months. Which of the following is closest to the total amount of their payments?

    a. $220
    b. $1,800
    c. $2,200
    d. $3,600

23. There are 1,000 meters in one kilometer. How many meters are there in 23 kilometers?

24. If the average adult sleeps 7 hours a night, how many hours of sleep does he or she get in one year? (1 year = 365 days)

25. Before he took a speed reading course, Joaquin read 250 words per minute. When he finished the course, he was able to read 375 words per minute. By how many words per minute did his reading speed increase?

Multiplication Review Chart

If you missed more than one problem on any group below, review the practice pages for those problems. Then redo the problems you got wrong before going on to the Division Skills Inventory. If you had a passing score, redo any problem you missed and begin the Division Skills Inventory on page 87.

PROBLEM NUMBERS	SKILL AREA	PRACTICE PAGES
1, 2	place value	6–12
3, 4	addition	15–30
5, 6, 25	subtraction	35–56
7, 8, 9, 10	simple multiplication	62–70
11, 12, 13, 14, 15	multiplying and carrying	71–73
16, 17, 18	multiplying by 10, 100, 1000	74–75
19, 20, 21	rounding and estimating	76–78
22, 23, 24	applying multiplication	79–83

DIVISION

Division Skills Inventory

Do all problems that you can. There is no time limit. Work carefully and check your answers, but do not use outside help.

1. $8\overline{)184}$

2. $4\overline{)268}$

3. $9\overline{)2,142}$

4. $7\overline{)2,931}$

5. $6\overline{)4,225}$

6. $5\overline{)24,014}$

7. $371 \div 53 =$

8. $331 \div 62 =$

9. $\dfrac{432}{48} =$

10. $\dfrac{2,656}{32} =$

11. $\dfrac{6,663}{73} =$

12. $51,801 \div 83 =$

13. $24,882 / 429 =$

14. $40,837 \div 583 =$

15. $\dfrac{5,600}{80} =$

16. Round 2,763 to the nearest *hundred.* Then divide by 7.

    $2{,}763 \div 7$

17. Round 18,274 to the nearest *thousand.* Then divide by 6.

    $18{,}274 \div 6$

18. Which of the following is *closest* to the correct answer for $22{,}698 \div 39$?

    a. 390
    b. 450
    c. 580
    d. 630

19. If a train travels at an average speed of 37 miles per hour, how much time will it take to go 814 miles?

20. A large carton weights 2,608 ounces. There are 16 ounces in one pound. Find the weight of the carton in pounds.

21. The Martinez family pays $5,616 in rent in one year. How much rent do they pay each month?

22. Petra bought four pairs of boots for her children. She paid $119.20 for all the boots. Find the average price for each pair of boots.

Division Skills Inventory Chart

If you missed more than one problem on any group below, work through the practice pages for that group. Then redo the problems you got wrong on the Division Skills Inventory. If you had a passing score on all seven groups of problems, redo any problem you missed and go to Posttest A on page 120 or Posttest B on page 125.

PROBLEM NUMBERS	SKILL AREA	PRACTICE PAGES
1, 2, 3	dividing by 1 digit	93–95
4, 5, 6	dividing with remainders	96–98
7, 8, 9, 10, 11, 12	dividing by 2 digits	100–104
13, 14	dividing by 3 digits	105–106
15, 16, 17	rounding and estimating	107–108
18	2-digit accuracy	109–110
19, 20, 21, 22	applying division	111–116

Basic Division Facts

This page and the following page will help you learn the basic division facts. These are more examples of **mental math.** The division facts are basic building blocks for further study of mathematics. You must *memorize* these facts. Check your answers to this exercise. Practice the facts you got wrong. Then try these problems again until you can get all the answers quickly and accurately. If you have trouble, go back to the multiplication table on page 65 for review.

Parts of a Division Problem

There are four common ways to write division. The number being divided is called the **dividend.** The number that divides into the dividend is the **divisor.** The answer is the **quotient.** In each example below, the dividend is 20, the divisor is 4, and the quotient is 5.

$$4{\overline{)20}} \quad\quad 20 \div 4 = 5 \quad\quad \frac{20}{4} = 5 \quad\quad 20 / 4 = 5$$

1. $27 \div 9 =$ $\quad\quad\quad\quad$ $54 \div 6 =$ $\quad\quad\quad\quad$ $25 \div 5 =$

2. $42 \div 6 =$ $\quad\quad\quad\quad$ $24 \div 8 =$ $\quad\quad\quad\quad$ $48 \div 8 =$

3. $16 \div 4 =$ $\quad\quad\quad\quad$ $45 \div 5 =$ $\quad\quad\quad\quad$ $12 \div 4 =$

4. $\dfrac{48}{6} =$ $\quad\quad\quad\quad$ $\dfrac{64}{8} =$ $\quad\quad\quad\quad$ $\dfrac{30}{5} =$

5. $\dfrac{18}{2} =$ $\quad\quad\quad\quad$ $\dfrac{108}{12} =$ $\quad\quad\quad\quad$ $\dfrac{32}{4} =$

6. $\dfrac{20}{5} =$ $\quad\quad\quad\quad$ $\dfrac{49}{7} =$ $\quad\quad\quad\quad$ $\dfrac{18}{6} =$

7. $81 / 9 =$ $\quad\quad\quad\quad$ $63 / 9 =$ $\quad\quad\quad\quad$ $56 / 8 =$

8. $36 / 6 =$ $\quad\quad\quad\quad$ $72 / 12 =$ $\quad\quad\quad\quad$ $21 / 7 =$

9. $56 / 7 =$ $\quad\quad\quad\quad$ $88 / 11 =$ $\quad\quad\quad\quad$ $54 / 9 =$

10. $15 \div 3 =$ $80 \div 10 =$ $40 \div 5 =$

11. $36 \div 4 =$ $42 \div 7 =$ $24 \div 6 =$

12. $72 \div 8 =$ $28 \div 7 =$ $35 \div 5 =$

13. $\dfrac{63}{7} =$ $\dfrac{50}{5} =$ $\dfrac{24}{3} =$

14. $\dfrac{16}{8} =$ $\dfrac{22}{2} =$ $\dfrac{32}{8} =$

15. $\dfrac{55}{11} =$ $\dfrac{40}{8} =$ $\dfrac{18}{3} =$

16. $60 / 12 =$ $0 / 2 =$ $45 / 9 =$

17. $4 / 1 =$ $21 / 3 =$ $60 / 5 =$

18. $30 / 6 =$ $90 / 9 =$ $24 / 12 =$

19. $84 \div 7 =$ $33 \div 11 =$ $14 \div 2 =$

20. $6 \div 3 =$ $24 \div 4 =$ $35 \div 7 =$

21. $72 \div 6 =$ $8 \div 1 =$ $9 \div 3 =$

22. $\dfrac{16}{2} =$ $\dfrac{96}{8} =$ $\dfrac{10}{5} =$

23. $\dfrac{48}{12} =$ $\dfrac{14}{7} =$ $\dfrac{12}{2} =$

24. $\dfrac{18}{9} =$ $\dfrac{12}{3} =$ $\dfrac{84}{12} =$

25. $8 / 4 =$ $28 / 4 =$ $9 / 1 =$

26. $27 / 3 =$ $0 / 6 =$ $88 / 8 =$

27. $36 / 9 =$ $30 / 10 =$ $8 / 2 =$

28. $15 \div 5 =$ $121 \div 11 =$ $60 \div 6 =$

29. $50 \div 10 =$ $40 \div 4 =$ $96 \div 12 =$

30. $7 \div 1 =$ $72 \div 9 =$ $0 \div 8 =$

31. $\dfrac{20}{5} =$ $\dfrac{66}{6} =$ $\dfrac{90}{10} =$

32. $\dfrac{22}{11} =$ $\dfrac{6}{1} =$ $\dfrac{8}{2} =$

33. $\dfrac{36}{3} =$ $\dfrac{108}{9} =$ $\dfrac{100}{10} =$

34. $40 / 10 =$ $36 / 12 =$ $33 / 3 =$

35. $20 / 10 =$ $132 / 12 =$ $44 / 4 =$

36. $6 / 2 =$ $12 / 6 =$ $66 / 11 =$

37. $99 \div 11 =$ $20 \div 2 =$ $60 \div 10 =$

38. $110 \div 11 =$ $70 \div 10 =$ $24 \div 2 =$

39. $144 \div 12 =$ $77 \div 11 =$ $10 \div 2 =$

Dividing by One-Digit Numbers

Look carefully at the example below to see how to do a division problem.

EXAMPLE

$$4\overline{)156}$$ with 3 above

STEP 1 Divide $15 \div 4 = 3$. Put the 3 over the 5 in the dividend.

$$4\overline{)156}$$ with 3 above, 12 below

STEP 2 Multiply $3 \times 4 = 12$. Put the 12 directly under the 15.

$$4\overline{)156}$$ with 3 above, 12 below, 3 remainder

STEP 3 Subtract $15 - 12 = 3$.

$$4\overline{)156}$$ with 3 above, 12 below, 36 brought down

STEP 4 Bring down the next number from the dividend.

$$4\overline{)156}$$ with 39 above, 12 below, 36

STEP 5 Divide $36 \div 4 = 9$. Put the 9 directly over the 6 in the dividend.

$$4\overline{)156}$$ with 39 above, 12, 36, 36

STEP 6 Multiply $9 \times 4 = 36$. Put the 36 directly under the 36 in the problem.

$$4\overline{)156}$$ with 39 above, 12, 36, 36, 0

STEP 7 Subtract $36 - 36 = 0$.

When dividing by a 1-digit number, most of the work can be done by multiplying and subtracting mentally. Write each digit that you carry in the dividend. This procedure is called **short division**.

EXAMPLES

$$4\overline{)15\overset{3}{6}} = 39 \qquad 5\overline{)1,3\overset{3}{7}\overset{2}{0}} = 274 \qquad 2\overline{)9,\overset{1}{7}3\overset{1}{4}} = 4,867$$

To check a division problem, multiply the quotient by the divisor. The product should be the dividend. For the examples above

$$\begin{array}{r} 39 \\ \times\ 4 \\ \hline 156 \end{array} \qquad \begin{array}{r} 274 \\ \times\ 5 \\ \hline 1,370 \end{array} \qquad \begin{array}{r} 4,867 \\ \times\ 2 \\ \hline 9,734 \end{array}$$

....

Divide and check.

1. $6\overline{)252}$ $\qquad$ $4\overline{)356}$ $\qquad$ $8\overline{)424}$ $\qquad$ $7\overline{)322}$ $\qquad$ $3\overline{)237}$

2. $3\overline{)162}$ $\qquad$ $9\overline{)801}$ $\qquad$ $2\overline{)126}$ $\qquad$ $6\overline{)504}$ $\qquad$ $8\overline{)688}$

3. $5\overline{)215}$ $\qquad$ $6\overline{)408}$ $\qquad$ $3\overline{)195}$ $\qquad$ $9\overline{)603}$ $\qquad$ $2\overline{)194}$

4. $8\overline{)272}$ $\qquad$ $5\overline{)360}$ $\qquad$ $4\overline{)264}$ $\qquad$ $7\overline{)336}$ $\qquad$ $6\overline{)174}$

5. $5\overline{)340}$ $\qquad$ $4\overline{)192}$ $\qquad$ $7\overline{)413}$ $\qquad$ $3\overline{)261}$ $\qquad$ $9\overline{)540}$

Rewrite each problem and divide. The first problem is started for you.

6. $360 \div 4 =$ $480 \div 8 =$ $240 \div 6 =$ $210 \div 3 =$ $450 \div 5 =$

$4\overline{)360}$

7. $2,768 \div 8 =$ $2,051 \div 7 =$ $1,940 \div 5 =$ $4,068 \div 9 =$

8. $2,325 \div 3 =$ $4,571 \div 7 =$ $2,065 \div 5 =$ $5,728 \div 8 =$

9. $\dfrac{3,624}{6} =$ $\dfrac{1,228}{4} =$ $\dfrac{7,232}{8} =$ $\dfrac{3,040}{5} =$

10. $1,418 / 2 =$ $3,448 / 8 =$ $2,468 / 4 =$ $3,432 / 6 =$

11. $\dfrac{3,420}{9} =$ $\dfrac{3,305}{5} =$ $\dfrac{2,184}{7} =$ $\dfrac{2,032}{4} =$

12. $2,624 / 4 =$ $1,746 / 6 =$ $1,827 / 9 =$ $3,752 / 7 =$

13. $\dfrac{2,220}{3} =$ $\dfrac{1,768}{8} =$ $\dfrac{2,370}{2} =$ $\dfrac{8,874}{9} =$

Dividing with Remainders

Division problems do not always come out evenly. The amount left over is called the **remainder.**

To check a division problem with a remainder, multiply the quotient by the divisor. Then add the remainder. The result should equal the dividend.

EXAMPLE

$$\begin{array}{r} 487 \text{ r } 4 \\ 5\overline{)2,439} \\ \underline{2\ 0} \\ 43 \\ \underline{40} \\ 39 \\ \underline{35} \\ 4 \end{array}$$

CHECK

$$\begin{array}{r} 487 \\ \times\quad 5 \\ \hline 2,435 \\ +\quad 4 \\ \hline 2,439 \end{array}$$

Divide and check.

1. $6\overline{)1,449}$ $8\overline{)5,629}$ $5\overline{)2,352}$ $7\overline{)2,059}$

2. $4\overline{)2,431}$ $6\overline{)2,495}$ $3\overline{)1,537}$ $9\overline{)8,197}$

3. $2\overline{)1,619}$ $5\overline{)1,842}$ $7\overline{)2,946}$ $4\overline{)2,867}$

Rewrite each problem. Then divide and check.

4. $4,295 \div 8 =$ $2,835 \div 4 =$ $5,954 \div 9 =$ $1,259 \div 3 =$

5. $1,321 \div 7 =$ $1,618 \div 6 =$ $2,163 \div 5 =$ $6,014 \div 8 =$

6. $1,623 \div 9 =$ $3,508 \div 8 =$ $1,846 \div 6 =$ $3,522 \div 4 =$

7. $18,834 \div 8 =$ $15,653 \div 4 =$ $15,680 \div 3 =$ $43,134 \div 5 =$

8. $58,810 \div 7 =$ $24,404 \div 6 =$ $74,163 \div 9 =$ $14,235 \div 2 =$

9. $\dfrac{8,335}{4} =$ $\dfrac{20,299}{5} =$ $\dfrac{52,022}{8} =$ $\dfrac{12,503}{6} =$

10. $\dfrac{15,014}{3} =$ $\dfrac{21,755}{7} =$ $\dfrac{14,057}{2} =$ $\dfrac{19,203}{4} =$

11. $\dfrac{54,725}{6} =$ $\dfrac{54,727}{9} =$ $\dfrac{40,053}{8} =$ $\dfrac{21,308}{3} =$

12. $\dfrac{5,530}{9} =$ $\dfrac{68,423}{7} =$ $\dfrac{83,212}{3} =$ $\dfrac{11,452}{6} =$

13. $21,815 / 8 =$ $98,167 / 5 =$ $62,189 / 4 =$ $44,981 / 7 =$

14. $78,424 / 9 =$ $12,893 / 8 =$ $61,254 / 7 =$ $83,481 / 6 =$

15. $12,290 / 3 =$ $19,538 / 5 =$ $38,242 / 4 =$ $22,517 / 8 =$

Mental Math and Properties of Numbers

Before you tackle dividing by 2-digit numbers, take the time to review all the basic operations with zeros and ones.

 Do each of the following in your head.

1. $8 \div 1 =$ 2. $1 + 6 =$ 3. $5 - 1 =$ 4. $9 + 0 =$

5. $7 \times 1 =$ 6. $4 + 0 =$ 7. $8 - 0 =$ 8. $9 \times 1 =$

9. $0 \div 7 =$ 10. $0 \times 4 =$ 11. $1 - 0 =$ 12. $5 \div 1 =$

13. $3 \times 0 =$ 14. $6 + 1 =$ 15. $1 \div 1 =$ 16. $1 \times 7 =$

17. $0 + 8 =$ 18. $1 \times 0 =$ 19. $9 - 1 =$ 20. $0 \div 3 =$

Be sure your answers are correct. Then read how these problems illustrate several properties about the basic operations.

For example, look at problems 2 and 14: $1 + 6 = 7$ and $6 + 1 = 7$. These two problems are examples of the **commutative property of addition.** In simple terms, the numbers in an addition problem can be added in any order.

Look at problems 5 and 16: $7 \times 1 = 7$ and $1 \times 7 = 7$. These two problems are examples of the **commutative property of multiplication.** The numbers in a multiplication problem can be multiplied in any order.

However, subtraction and division are *not* commutative. In problem 3, $5 - 1 = 4$, but $1 - 5$ is an algebra problem involving negative numbers. In problem 1, $8 \div 1 = 8$, but $1 \div 8$ is a fraction or decimal problem.

Look at problems 5, 8, and 16. These problems illustrate the fact that **any number multiplied by one is that number.**

Look at problems 1, 12, and 15. These problems illustrate the fact that **any number divided by one is that number.**

Look at problems 10, 13, and 18. These problems illustrate the fact that **any number multiplied by zero is zero.**

Look at problems 9 and 20. These problems illustrate the fact that **any number divided into zero is zero.** Notice that there is no problem such as $8 \div 0$. There is no number to multiply by zero to get 8. You cannot divide a number by zero. Mathematicians say that $8 \div 0$ is **undefined.**

Dividing by Two-Digit Numbers

Dividing by two-digit and three-digit numbers is a tricky process. It requires practice and a skill called **estimating;** that is, guessing how many times one number goes into another. Look at these examples carefully.

EXAMPLE 1

$$\begin{array}{r} 46 \\ 32\overline{)1{,}472} \\ 1\,28 \\ \hline 192 \\ 192 \end{array}$$

STEP 1 Ask yourself how many times 32 goes into 147. To estimate, ask how many times 3 goes into 14. $14 \div 3 = 4$ with a remainder.

STEP 2 Place the 4 over the 7 and multiply $4 \times 32 = 128$.

STEP 3 Subtract $147 - 128 = 19$.

STEP 4 Bring down the 2.

STEP 5 Ask yourself how many times 32 goes into 192. To estimate, ask how many times 3 goes into 19. $19 \div 3 = 6$ with a remainder.

STEP 6 Place the 6 over the 2 and multiply $6 \times 32 = 192$.

STEP 7 Subtract $192 - 192 = 0$.

CHECK

$$\begin{array}{r} 46 \\ \times\ 32 \\ \hline 92 \\ 1\,38 \\ \hline 1{,}472 \end{array}$$

STEP 8 Check $32 \times 46 = 1{,}472$.

EXAMPLE 2

$$\begin{array}{r} 58 \\ 47\overline{)2{,}726} \\ 2\,35 \\ \hline 376 \\ 376 \end{array}$$

STEP 1 Ask yourself how many times 47 goes into 272. To estimate, ask how many times 4 goes into 27. $27 \div 4 = 6$ with a remainder.

STEP 2 Place the 6 over the 2 and multiply $6 \times 47 = 282$. Since 282 is bigger than 272, and therefore can't be subtracted from 272, we know that the 6 is too big. Erase the 6 and try 5. Multiply $5 \times 47 = 235$.

STEP 3 Subtract $272 - 235 = 37$.

STEP 4 Bring down the 6.

STEP 5 Ask yourself how many times 47 goes into 376. To estimate, ask how many times 4 goes into 37. $37 \div 4 = 9$ with a remainder.

STEP 6 Place the 9 over the 6 and multiply $9 \times 47 = 423$. Since 423 is bigger than 376, and therefore can't be subtracted from 376, we know that the 9 is too big. Erase the 9 and try 8. Multiply $8 \times 47 = 376$.

STEP 7 Subtract $376 - 376 = 0$.

STEP 8 Check $47 \times 58 = 2,726$.

CHECK

$$
\begin{array}{r}
58 \\
\times\ 47 \\
\hline
406 \\
2\ 32 \\
\hline
2,726
\end{array}
$$

....

Divide and check.

1. $42\overline{)336}$ $\qquad$ $23\overline{)161}$ $\qquad$ $62\overline{)310}$ $\qquad$ $83\overline{)332}$

2. $52\overline{)312}$ $\qquad$ $63\overline{)504}$ $\qquad$ $72\overline{)288}$ $\qquad$ $92\overline{)828}$

3. $19\overline{)133}$ $\qquad$ $48\overline{)288}$ $\qquad$ $57\overline{)228}$ $\qquad$ $69\overline{)207}$

4. $87\overline{)174}$ $\qquad$ $79\overline{)395}$ $\qquad$ $38\overline{)228}$ $\qquad$ $47\overline{)329}$

5. $22\overline{)176}$ $56\overline{)168}$ $63\overline{)252}$ $75\overline{)450}$

Rewrite each problem. Then divide and check.

6. $465 \div 93 =$ $432 \div 48 =$ $162 \div 54 =$ $460 \div 46 =$

7. $217 \div 31 =$ $108 \div 18 =$ $68 \div 17 =$ $672 \div 96 =$

8. $200 \div 25 =$ $320 \div 64 =$ $342 \div 38 =$ $318 \div 53 =$

9. $683 \div 82 =$ $283 \div 43 =$ $266 \div 65 =$ $132 \div 39 =$

10. $538 \div 74 =$ $271 \div 52 =$ $344 \div 38 =$ $150 \div 41 =$

11. $\dfrac{136}{29} =$ $\dfrac{66}{18} =$ $\dfrac{275}{43} =$ $\dfrac{207}{27} =$

12. $\dfrac{261}{46} =$ $\dfrac{353}{87} =$ $\dfrac{206}{52} =$ $\dfrac{598}{74} =$

13. $\dfrac{221}{66} =$ $\dfrac{191}{35} =$ $\dfrac{246}{56} =$ $\dfrac{201}{21} =$

14. $\dfrac{736}{32} =$ $\dfrac{1,845}{41} =$ $\dfrac{3,286}{53} =$ $\dfrac{5,022}{62} =$

15. $\dfrac{4,648}{83} =$ $\dfrac{3,312}{72} =$ $\dfrac{836}{22} =$ $\dfrac{2,484}{92} =$

16. 3,496 / 38 =   3,087 / 49 =   2,394 / 57 =   4,828 / 68 =

17. 2,352 / 98 =   2,736 / 76 =   667 / 29 =   3,696 / 88 =

18. 693 / 43 =   624 / 28 =   1,014 / 72 =   2,599 / 81 =

19. $\dfrac{768}{19} =$ $\dfrac{1,687}{56} =$ $\dfrac{1,859}{37} =$ $\dfrac{1,742}{29} =$

20. $\dfrac{4,222}{52} =$ $\qquad$ $\dfrac{2,666}{63} =$ $\qquad$ $\dfrac{2,256}{44} =$ $\qquad$ $\dfrac{3,166}{98} =$

21. $\dfrac{1,793}{76} =$ $\qquad$ $\dfrac{1,894}{28} =$ $\qquad$ $\dfrac{955}{55} =$ $\qquad$ $\dfrac{1,670}{63} =$

22. $\dfrac{1,048}{20} =$ $\qquad$ $\dfrac{1,650}{40} =$ $\qquad$ $\dfrac{4,252}{80} =$ $\qquad$ $\dfrac{567}{50} =$

23. $12,168 \div 52 =$ $\qquad$ $13,546 \div 26 =$ $\qquad$ $46,209 \div 73 =$ $\qquad$ $20,482 \div 49 =$

24. $20,368 \div 38 =$ $\qquad$ $14,814 \div 18 =$ $\qquad$ $25,480 \div 56 =$ $\qquad$ $40,256 \div 64 =$

25. $\dfrac{29,376}{72} =$ $\qquad$ $\dfrac{44,616}{88} =$ $\qquad$ $\dfrac{12,669}{41} =$ $\qquad$ $\dfrac{32,224}{53} =$

26. $\dfrac{53,460}{99} =$ $\qquad$ $\dfrac{17,280}{54} =$ $\qquad$ $\dfrac{21,280}{28} =$ $\qquad$ $\dfrac{18,130}{37} =$

Dividing by Three-Digit Numbers

EXAMPLE

$$632\overline{)27{,}176} \quad \begin{array}{r} 43 \\ \end{array}$$

$$\begin{array}{r} 25\ 28 \\ \hline 1\ 896 \\ 1\ 896 \\ \hline \end{array}$$

STEP 1 Ask yourself how many times 632 goes into 2717 (you know it won't go into 271). To estimate, ask how many times 6 goes into 27.
$27 \div 6 = 4$ with a remainder.

STEP 2 Place the 4 over the 7 and multiply $4 \times 632 = 2528$.

STEP 3 Subtract $2717 - 2528 = 189$.

STEP 4 Bring down the 6.

STEP 5 Ask yourself how many times 632 goes into 1896. To estimate, ask how many times 6 goes into 18. $18 \div 6 = 3$

STEP 6 Place the 3 over the 6 and multiply $3 \times 632 = 1896$.

STEP 7 Subtract $1896 - 1896 = 0$.

CHECK

$$\begin{array}{r} 632 \\ \times\ 43 \\ \hline 1\ 896 \\ 25\ 28 \\ \hline 27{,}176 \\ \end{array}$$

STEP 8 Check $632 \times 43 = 27{,}176$.

Divide and check.

1. $326\overline{)7{,}824}$ $418\overline{)15{,}884}$ $521\overline{)26{,}571}$ $607\overline{)38{,}241}$

2. $862\overline{)16{,}022}$ $793\overline{)17{,}867}$ $486\overline{)22{,}566}$ $972\overline{)33{,}086}$

Rewrite each problem. Then divide and check.

3. $\dfrac{28,504}{509} =$ $\dfrac{30,456}{423} =$ $\dfrac{54,531}{657} =$ $\dfrac{23,867}{823} =$

4. $\dfrac{34,704}{964} =$ $\dfrac{79,807}{877} =$ $\dfrac{18,792}{216} =$ $\dfrac{17,490}{330} =$

5. $16,739 \div 418 =$ $37,712 \div 628 =$ $63,183 \div 902 =$ $22,606 \div 451 =$

6. $37,659 \div 523 =$ $12,244 \div 139 =$ $25,611 \div 711 =$ $16,987 \div 269 =$

7. $16,692 / 321 =$ $37,400 / 425 =$ $41,406 / 618 =$ $24,381 / 903 =$

Rounding and Estimating

Not all division problems are as difficult as those in the last exercise. Numbers that end in zeros are often easy to work with.

EXAMPLE $\frac{3,600}{9} = 400$

STEP 1 Forget about the zeros in 3,600, and divide $36 \div 9 = 4$.

STEP 2 Bring along the zeros from 3,600.

 Divide each problem in your head.

1. $1,200 \div 4 =$ $1,600 \div 8 =$ $2,100 \div 7 =$

2. $2,000 \div 5 =$ $1,400 \div 2 =$ $40,000 \div 4 =$

3. $480 \div 6 =$ $3,300 \div 3 =$ $7,200 \div 9 =$

4. $630 \div 7 =$ $15,000 \div 5 =$ $30,000 \div 6 =$

When both the dividend and the divisor end in zeros, you can *cancel* the zeros one-for-one.

EXAMPLE $\frac{7,200}{80} = 90$

STEP 1 Forget about the zeros in 7,200 and 80, and divide $72 \div 8 = 9$.

STEP 2 Cancel the zeros one-for-one, and bring along the remaining zero.

 Divide each problem in your head.

5. $240 \div 40 =$ $8,100 \div 90 =$ $45,000 \div 90 =$

6. $18,000 \div 200 =$ $1,800 \div 30 =$ $200 \div 20 =$

7. $2,400 \div 600 =$ $4,900 \div 70 =$ $3,500 \div 70 =$

8. $15,000 \div 50 =$ $2,400 \div 300 =$ $64,000 \div 80 =$

Sometimes you can round the dividend to a number that is easy to divide into. You can use the rounded number to estimate an answer to the original division problem.

__EXAMPLE 1__ Estimate the answer to the problem $539 \div 9$.

$539 \div 9 \approx 540 \div 9$	__STEP 1__	Round 539 to the nearest ten.
$539 \div 9 \approx 500 \div 9$	__STEP 2__	Round 539 to the nearest hundred.
$540 \div 9 = 60$	__STEP 3__	Since 540 divides evenly by 9, use the problem $540 \div 9$ to estimate the answer.

__EXAMPLE 2__ Estimate the answer to the problem $209 \div 4$.

$209 \div 4 \approx 210 \div 4$	__STEP 1__	Round 209 to the nearest ten.
$209 \div 4 \approx 200 \div 4$	__STEP 2__	Round 209 to the nearest hundred.
$200 \div 4 = 50$	__STEP 3__	Since 200 divides evenly by 4, use the problem $200 \div 4$ to estimate the answer.

....

Round each dividend to the nearest *ten* and the nearest *hundred*. Then decide which number is easier to divide. Use the easier rounded number to estimate the answer.

9. $423 \div 7$ $409 \div 4$

10. $329 \div 6$ $492 \div 7$

11. $194 \div 5$ $958 \div 8$

12. $212 \div 3$ $716 \div 9$

Round each dividend to the nearest *hundred* and the nearest *thousand*. Then decide which rounded number is easier to divide. Use the easier rounded number to estimate the answer.

13. $3,186 \div 4$ $4,236 \div 8$

14. $3,186 \div 6$ $2,095 \div 7$

15. $8,077 \div 9$ $3,182 \div 5$

16. $1,790 \div 3$ $4,773 \div 6$

Two-Digit Accuracy and Using a Calculator

Another way to estimate answers is to do a partial division. Instead of completing a division problem, try for **two-digit accuracy.** Divide until you have three digits in the quotient. Then round your answer. Study the example carefully.

EXAMPLE Estimate an answer to the problem 158,916 ÷ 41.

$$
\begin{array}{r}
3,87 - \approx 3,900 \\
41\overline{)158,916} \\
\underline{123} \\
35\ 9 \\
\underline{32\ 8} \\
3\ 11
\end{array}
$$

STEP 1 Ask yourself how many times 41 goes into 158. To estimate, ask how many times 4 goes into 15. 15 ÷ 4 = 3 with a remainder.

STEP 2 Place 3 over the 8 and multiply 3 ÷ 31 = 123.

STEP 3 Subtract 158 – 123 = 35.

STEP 4 Bring down the 9.

STEP 5 Ask yourself how many times 41 goes into 359. To estimate, ask how many times 4 goes into 35. 35 ÷ 4 = 8 with a remainder.

STEP 6 Place 8 over the 9 and multiply 8 × 41 = 328.

STEP 7 Subtract 359 – 328 = 31.

STEP 8 Bring down the 1. Then decide whether 41 goes into 311 5 times or more. 31 ÷ 4 = 7 with a remainder. Since the digit to the right of 8 is 5 or more, raise 8 to 9, and put zeros in the tens and units places.

Using a Calculator To check the problem 158,916 ÷ 41 on a calculator, press the following keys:

The calculator display should read ⬚ 3876.

Calculate each problem to two-digit accuracy. Then check each problem with a calculator.

1. $\dfrac{7,936}{62} =$ $\dfrac{18,316}{76} =$ $\dfrac{14,798}{49} =$ $\dfrac{34,506}{81} =$

2. $\dfrac{20,909}{29} =$ $\dfrac{45,724}{92} =$ $\dfrac{18,444}{53} =$ $\dfrac{23,834}{34} =$

3. $\dfrac{105,705}{87} =$ $\dfrac{157,212}{66} =$ $\dfrac{144,648}{41} =$ $\dfrac{569,439}{93} =$

Use your knowledge of two-digit accuracy to select the correct answer from the choices.

4. $5,943 \div 21 =$

a. 83
b. 183
c. 213
d. 283

5. $23,072 \div 32 =$

a. 961
b. 721
c. 551
d. 341

6. $22,684 \div 53 =$

a. 618
b. 538
c. 428
d. 298

7. $19,034 \div 62 =$

a. 417
b. 397
c. 307
d. 267

8. $65,807 \div 79 =$

a. 833
b. 763
c. 703
d. 683

9. $13,708 \div 46 =$

a. 488
b. 438
c. 378
d. 298

10. $154,686 \div 58 =$

a. 1,927
b. 2,667
c. 3,107
d. 3,577

11. $365,556 \div 82 =$

a. 3,998
b. 4,108
c. 4,458
d. 5,018

12. $261,501 \div 67 =$

a. 2,543
b. 2,933
c. 3,023
d. 3,903

13. $448,622 \div 82 =$

a. 5,471
b. 5,991
c. 6,301
d. 6,781

14. $213,014 \div 73 =$

a. 3,388
b. 2,918
c. 2,278
d. 1,858

15. $108,824 \div 61 =$

a. 994
b. 1,254
c. 1,784
d. 2,164

Applying Your Division Skills

On this page and the following five pages are practical problems that require you to apply your division skills. In each problem, pay close attention to the language that tells you to divide.

You may be asked how many of one thing are contained in something else.

EXAMPLE 1   A truck can carry 2,000 pounds. How many boxes, each weighing 50 pounds, can the truck carry?

$$\frac{2,00\cancel{0}}{5\cancel{0}} = 40 \text{ boxes}$$ Find how many times 50 divides into 2,000.

You may be given information about several things and asked for information about one of those things.

EXAMPLE 2   Hector makes $84 for working 7 hours for a moving company. How much did he make each hour?

$$\frac{\$84}{7} = \$12$$ Divide the total wages, $84, by the number of hours he worked, 7.

You may be asked to find an average. An **average,** or **mean,** is a total number divided by the number of things that make up the total.

EXAMPLE 3   Together, three children weigh 201 pounds. What is the average weight of each child?

number of children $\longrightarrow$ $3\overline{)201}$ $\begin{matrix} 67 \longleftarrow \text{average weight} \\ \longleftarrow \text{total weight} \end{matrix}$ Divide 201 by 3.

Solve and write the correct label, such as $, miles, or pounds, next to each answer.

1. If 3 pounds of roast beef cost $17.97, find the price of 1 pound of beef.

2. There are 12 inches in 1 foot. Find the length in feet of a workbench that is 180 inches long.

3. There are 3 feet in 1 yard. Find the length in yards of the workbench in problem 2.

4. Doreen saved $1,976 in 1 year. What was the average amount she saved each week? (1 year = 52 weeks)

5. On a regional map, 1 inch is equal to 35 miles. How many inches apart on the map are two cities that are actually 455 miles apart?

6. A salesman told Audrey that she can pay only $24 a month for the furniture she wants. How many months will she need to pay for furniture that costs $912?

7. The total rain one year in San Juan, Puerto Rico, was 60 inches. What was the average monthly rainfall?

8. A paint store owner has 1,148 quarts of red paint. To save space, he wants to transfer the paint to gallon containers. There are 4 quarts in 1 gallon. How many gallon containers will he need?

9. If a dozen kitchen chairs cost $354, what is the cost of one chair?

10. A packer in the shipping department of a book company can put 36 copies of a new book in one carton. If she has 27,072 books to pack, how many cartons will she need?

11. Tim spent $2,664.50 on gasoline last year. What was the average amount he spent each day? (1 year = 365 days)

12. Manny bought a used car. He will make payments of $85 a month for the car. How many months will it take him to pay off a $2,720 car loan?

13. If a plane flies at an average speed of 436 miles per hour, how long will it take to fly 1,744 miles?

14. The Dean family spent $447.36 last year for gas and electricity. What was their average monthly gas and electricity bill?

15. An airplane has a cruising altitude of 36,960 feet above sea level. There are 5,280 feet in 1 mile. The cruising altitude of the plane is how many miles above sea level?

16. Mary wants to put tiles on the floor of her son's bedroom. She measured the floor space and found that the bedroom floor has 10,800 square inches. There are 144 square inches in 1 square foot. How many square feet of tiles does Mary need?

17. Rosa makes $22,672 in a year. How much does she make each week? (1 year = 52 weeks)

18. Paul took a speed reading class. He reads about 285 words per minute. Which of the following is closest to the number of minutes he needs to read an article that contains 12,000 words?

   a. 20 minutes
   b. 30 minutes
   c. 40 minutes
   d. 50 minutes

19. One shelf in Fernando's grocery will hold 16 cans of vegetables. If he has 512 cans, how many shelves can he fill?

20. The Soto family spends $4,596 a year for rent. What is their monthly rent?

21. An airplane travels at an average speed of 365 miles per hour. How many hours does the plane require to travel a distance of 5,110 miles?

22. The Jackson family drove a total of 5,376 miles on their vacation. If they traveled for 14 days, how many miles did they average each day?

23. A freight elevator will safely hold 3,600 pounds. If one crate weighs 150 pounds, how many crates can the elevator hold?

24. One season a professional basketball player scored 1,638 points. He played in 78 games that season. What was his average score per game?

25. There are 2,000 pounds in 1 ton. How many tons are there in 360,000 pounds?

26. The State Theater has 1,512 seats. If each row contains 36 seats, how many rows are there in the theater?

27. Mavis owes $333 for a sound system. She will have to make 18 equal monthly payments. How much will she pay each month?

28. A case containing a dozen quarts of motor oil costs $11.88. What is the price of 1 quart of motor oil?

29. The store that sells the motor oil in the last problem charges $1.29 for each quart of oil that it sells separately. How much more does it cost to buy a dozen quarts of motor oil separately than to buy a full case?

30. Sam works part-time for a tree trimming service. In 4 years he earned $46,000. Find the average amount he earned per year.

Division Review

This review covers the material you have studied so far in this book. Read the signs carefully to decide whether to add, subtract, multiply, or divide.

1. The factory where Casey works produced 13,826 machine parts in one month. Round the number of parts to the nearest *thousand*.

2. What is the value of the digit 8 in the number 13,826?

3. $236 + 4,807 =$

4. $26,458 + 9,583 =$

5. $16 + 56 + 384 =$

6. $8,134 - 2,023 =$

7. $25,126 - 19,587 =$

8. $250,000 - 83,056 =$

9. $6 \times 53 =$

10. $(24)(87) =$

11. $36 \cdot 258 =$

12. $(724)(1,000) =$

13. $258 \div 6 =$

14. $5,022 / 8 =$

15. $496 \div 62 =$

16. $\dfrac{3,096}{43} =$

17. $33,369 \div 49 =$

18. $20,808 \div 289 =$

19. Round both numbers to the nearest *hundred* and multiply.

(293)(719)

 20. Round the larger number to the nearest *thousand*. Then divide by 9.

71,648 ÷ 9

 21. A plane flew at an average speed of 415 miles per hour for 7 hours. How far did the plane travel?

 22. Once a month a maintenance crew has to wax the floors of the conference rooms in a large office. The conference rooms have a combined floor area of 7,800 square feet. Each can of wax will cover 600 square feet. How many cans of wax does the crew need to do the floors of the conference rooms?

23. Margaret earned $20,124 last year. How much did she earn each week? (52 weeks = 1 year)

24. Judy drove 989 miles on a trip to North Carolina. She had to buy 43 gallons of gasoline on the trip. How many miles per gallon did she average on the trip?

Division Review Chart

If you missed more than one problem on any group below, review the practice pages for those problems. Then redo the problems you got wrong before going on to Posttest A or B. If you had a passing score, redo any problem you missed and begin Posttest A on page 120 or Posttest B on page 125.

PROBLEM NUMBERS	SHILL AREA	PRACTICE PAGES
1, 2	place value	6–12
3, 4, 5	addition	15–30
6, 7, 8	subtraction	35–56
9, 10, 11, 12, 21	multiplication	62–83
13, 14	dividing by 1 digit	93–95
15, 16, 17, 18	dividing by 2 and 3 digits	100–106
19, 20	rounding and estimating	107–108
22, 23, 24	applying division	111–116

Posttest A

....

This posttest gives you a chance to check your basic mathematical skills. Take your time and work each problem carefully.

1. Which digit in the number 56,324 is in the thousands place?

2. What is the value of the digit 3 in 56,324?

3. The population of Madison County is 249,238. Round the number to the nearest *ten thousand.*

4. In a recent year, the state of Texas spent $3,692 per student in their public school system. Round $3,692 to the nearest *hundred* dollars.

5.　126
　　+ 543

6.　12,082
　　　9,565
　　+　 788

7. $9 + 84 + 71 + 6 =$

8. $125,036 + 84,591 =$

9. $76,053 + 248 + 1,592 =$

10. Round 2,724 to the nearest *thousand*. Round 809 to the nearest *hundred*. Round 19,628 to the nearest *ten thousand*. Then add.

 11. The driving distance from Atlanta to Washington, D.C., is 630 miles. The distance from Washington, D.C., to New York City is 229 miles. The distance from New York City to Boston is 216 miles. Find the total driving distance from Atlanta to Boston by way of Washington, D.C., and New York.

 12. Maria and Tom were recently married. From her savings, Maria will spend $5,600 as a down payment for a new house. Tom can contribute $3,500. Tom's parents promise to give the couple $8,000. Find the total amount Maria and Tom have for their down payment.

13.
$$
\begin{array}{r}
358 \\
-216 \\
\hline
\end{array}
$$

14.
$$
\begin{array}{r}
8,542 \\
-5,296 \\
\hline
\end{array}
$$

15. $2,643 - 1,978 =$

16. $300,000 - 78,140 =$

17. $460,200 - 381,530 =$

 18. Round each number to the nearest *thousand* and subtract.

$49,528 - 17,485$

19. The price of a popular compact car ranges from $11,455 for the basic model to $13,290 for the model with all the accessories. The basic model is how much less than the model with all the accessories?

 20. The Clarkville Central School hopes to raise $100,000 for scholarships. So far they have raised $83,965. How much more do they need to raise?

21. 23
   × 32

22. 1,984
   ×   7

23. 89 × 67 =

24. 1,000 • 382 =

25. (1,736)(73) =

 26. Round 48 to the nearest *ten* and 3,280 to the nearest *thousand*. Then multiply.

27. On a state map 1 inch represents a distance of 35 miles. Two towns are 5 inches apart on the map. What is the actual distance in miles between the two towns?

 28. The owner of a car dealership must pay a supplier $196 for a popular model of an AM/FM/CD player. How much will he pay the supplier for 15 of the players?

29. $6\overline{)468}$

30. $8\overline{)7,237}$

31. $4,067 \div 49 =$

32. $\dfrac{41,400}{90} =$

33. $16,132 / 218 =$

 34. Find the answer to the problem $395,884 \div 76$ to the nearest *thousand.*

 35. An auditorium has 1,610 seats. Every row in the auditorium has 35 seats. How many rows of seats are there in the auditorium?

36. A warehouse received a shipment that weighed a total of 3,723 pounds. The shipment consisted of identical boxes of small motor parts, each weighing 17 pounds. How many boxes were in the shipment?

Posttest A Prescriptions

Circle the number of any problem that you miss. If you missed one or less in each group below, go on to Using Number Power on page 129. If you missed more than one problem in any group, review the chapters in this book or refer to these practice pages in other materials from Contemporary Books.

PROBLEM NUMBERS	PRESCRIPTION MATERIALS	PRACTICE PAGES
1, 2, 3, 4	place value	6–12
	Breakthroughs in Math: Book 1	7–16
	Math Skills That Work: Book 1	2–22
5, 6, 7, 8, 9, 10, 11, 12	addition	15–30
	Breakthroughs in Math: Book 1	18–41
	Math Skills That Work: Book 1	28–55
	Real Numbers: Estimation 1	5–15
	Math Exercises: Whole Numbers and Money	3–9
13, 14, 15, 16, 17, 18, 19, 20	subtraction	35–56
	Breakthroughs in Math: Book 1	44–69
	Math Skills That Work: Book 1	60–87
	Real Numbers: Estimation 1	16–22
	Math Exercises: Whole Numbers and Money	10–15
21, 22, 23, 24, 25, 26, 27, 28	multiplication	62–83
	Breakthroughs in Math: Book 1	72–97
	Math Skills That Work: Book 1	92–125
	Real Numbers: Estimation 1	23–29
	Math Exercises: Whole Numbers and Money	17–21
29, 30, 31, 32, 33, 34, 35, 36	division	90–116
	Breakthroughs in Math: Book 1	100–131
	Math Skills That Work: Book 1	130–161
	Real Numbers: Estimation 1	30–36
	Math Exercises: Whole Numbers and Money	22–27

For further practice:

Math Solutions (software)
Whole Numbers

Posttest B

This posttest has a multiple-choice format much like many standardized tests. Take your time and work each problem carefully. Circle the correct answer to each problem.

1. Subtract 6,894 from 70,293.

   a. 63,399
   b. 64,109
   c. 64,799
   d. 64,979

2. What is the sum of 1,296 and 3,978?

   a. 6,154
   b. 5,274
   c. 5,184
   d. 4,934

3. The CN Tower in Toronto is 1,821 feet high. What is the value of the digit 8 in the number 1,821?

   a. 8
   b. 80
   c. 800
   d. 8000

4. Silvia's new sofa cost $783 including tax and delivery charges. She made a down payment of $135. Find the balance due on the sofa.

   a. $588
   b. $598
   c. $618
   d. $648

5. Silvia, in the last problem, paid off the balance in 12 equal monthly payments. How much did she pay each month?

   a. $44
   b. $48
   c. $52
   d. $54

6. In the number 82,564, which digit is in the thousands place?

   a. 8
   b. 2
   c. 5
   d. 6

7. $4,856 \div 8 =$

   a. 670
   b. 667
   c. 607
   d. 67

8. $204,070 - 8,329 =$

   a. 196,951
   b. 195,741
   c. 204,359
   d. 212,399

9. $487 \times 9 =$

   a. 3,893
   b. 3,913
   c. 4,383
   d. 4,423

10. $59 + 3 + 47 + 1 =$

    a. 100
    b. 110
    c. 120
    d. 130

11. $10,000 - 9,326 =$

   a. 1,034

   b. 994

   c. 784

   d. 674

12. Mr. and Mrs. Alvarez pay $677 a month for rent. How much rent do they pay in one year?

   a. $6,770

   b. $7,190

   c. $8,124

   d. $9,034

13. $544 / 32 =$

   a. 16

   b. 17

   c. 21

   d. 22

14. $\dfrac{8,235}{27} =$

   a. 350

   b. 335

   c. 325

   d. 305

15. $56 + 2,043 + 837 =$

   a. 2,936

   b. 4,126

   c. 8,440

   d. 16,013

16. $(74)(358) =$

   a. 26,492

   b. 25,942

   c. 24,392

   d. 21,082

17. $56,032 - 9,547 =$

   a. 65,579

   b. 55,479

   c. 46,485

   d. 45,395

18. $16 \cdot 4,039 =$

   a. 60,544

   b. 61,414

   c. 62,864

   d. 64,624

19. The balance due on Mr. and Mrs. Chan's mortgage is $59,927. Round the balance to the nearest *thousand* dollars.

   a. $57,000

   b. $58,000

   c. $59,000

   d. $60,000

20. Herbert works 4 days a week as a traveling salesman. On Monday he drove 417 miles, on Tuesday he drove 308 miles, on Wednesday he drove 289 miles, and on Thursday he drove 318 miles. Altogether, how many miles did he drive those days?

   a. 725

   b. 1,014

   c. 1,043

   d. 1,332

21. In the last problem, find the average number of miles that Herbert drove each day.

   a. 293

   b. 333

   c. 343

   d. 373

22. In a week Herbert often drives 1,500 miles. The total distance that he drove on the 4 days listed in problem 20 is how many miles less than 1,500?

   a. 32

   b. 68

   c. 132

   d. 168

23. 3,741 + 29,634 + 8,025 =

    a. 41,400
    b. 44,050
    c. 75,069
    d. 147,294

24. 28,700 ÷ 70 =

    a. 4,101
    b. 4,100
    c. 410
    d. 41

25. Round each number in the problem 8,672 + 17,386 to the nearest *thousand*. Then add.

    a. 24,000
    b. 25,000
    c. 26,000
    d. 27,000

26. In a recent year, the U.S. imported 245,296 passenger cars from Germany. Round the number of imported cars to the nearest *ten thousand*.

    a. 200,000
    b. 240,000
    c. 250,000
    d. 260,000

27. (293)(1,000) =

    a. 2,930
    b. 29,300
    c. 293,000
    d. 2,930,000

28. In a town election, three candidates ran for councilman. The winner received 1,284 votes. The second-place candidate received 996 votes, and the third-place candidate received 381 votes. What was the total number of votes for the position of councilman?

    a. 1,861
    b. 2,371
    c. 2,551
    d. 2,661

29. In the problem 12,827 − 8,549, round each number to the nearest *thousand*. Then subtract.

    a. 3,000
    b. 4,000
    c. 5,000
    d. 6,000

30. From her weekly check, Carmen's employer deducts $83.50 for federal tax, $22.34 for social security, and $19.72 for state tax. Find the total of these deductions.

    a. $145.46
    b. $125.56
    c. $115.64
    d. $105.84

31. Carmen, in the last problem, makes a gross salary of $480.00 per week. What is her weekly take-home pay?

    a. $354.44
    b. $325.56
    c. $296.44
    d. $286.56

32. What is the quotient of 49,634 divided by  13 to the nearest *hundred*?

    a. 800
    b. 3,600
    c. 3,700
    d. 3,800

33. Round both 904 and 7,308 to the nearest *thousand*. Then multiply.

    a. 70,000,000
    b. 63,000,000
    c. 7,000,000
    d. 6,300,000

34. Which of the following is the *least* amount that is sufficient to pay for 5 gallons of paint that cost $18.79 per gallon?

    a. $75
    b. $85
    c. $90
    d. $100

35. One month a travel agency sold 387 tickets for a total of $54,958. Which of the following is closest to the average price of one ticket?

    a. $90
    b. $140
    c. $190
    d. $220

Posttest B Chart

If you missed more than one problem on any group below, review the practice pages for those problems. Then redo the problems you got wrong before going on to Using Number Power. If you had a passing score, redo any problem you missed and go on to Using Number Power on page 129.

PROBLEM NUMBERS	SKILL AREA	PRACTICE PAGES
3, 6, 19, 26	place value	6–12
2, 10, 15, 20, 23, 25, 28, 30	addition	15–30
1, 4, 8, 11, 17, 22, 29, 31	subtraction	35–56
9, 12, 16, 18, 27, 33, 34	multiplication	62–83
5, 7, 13, 14, 21, 24, 32, 35	division	90–116

Using
Number
Power

Changing Units of Measurement

This page and the following four pages will give you practice working with units of measurement. To change from one unit of measurement to another, refer to the table on page 168. It is not necessary to memorize the table, but you should become familiar with the measures listed there.

To change **from a large unit** of measurement **to a small unit, multiply** by the number of small units contained in one large unit.

<u>EXAMPLE 1</u> Change 6 pounds to ounces.

> **STEP 1** Check the Table of Measurements on page 168 to find out how many of the small units are contained in one large unit. 1 pound = 16 ounces

> **STEP 2** Multiply $16 \times 6 = 96$ ounces.

To change **from a small unit** of measurement **to a large unit, divide** by the number of small units contained in one large unit.

<u>EXAMPLE 2</u> Change 48 inches to feet.

> **STEP 1** Check the Table of Measurements on page 168 to find out how many of the small units are contained in one large unit. 1 foot = 12 inches

> **STEP 2** Divide $\quad 12\overline{)48}^{\,4}$

Change each of the following to the unit indicated. Use the Table of Measurements to find out how many small units are contained in the large units in each problem.

1. 192 ounces = _____ pounds

2. 5 miles = _____ feet

3. 4,000 meters = _____ kilometers

4. 54 feet = _____ yards

5. 72 gallons = _____ quarts

6. 12 kilograms = _____ grams

7. 504 inches = _____ yards

8. 8 hours = _____ minutes

9. 425 meters = _____ centimeters

10. 148 pints = _____ quarts

11. 15,840 yards = _____ miles

12. 420 quarts = _____ gallons

Adding Measurements

EXAMPLE	2 ft 9 in.	STEP 1	Add each unit of measurement separately.
	3 ft 10 in.		
	+ 1 ft 8 in.		
	6 ft 27 in.		

$$\begin{array}{r} 2\text{ ft }3\text{ in.} \\ 12\overline{)27} \\ 24 \\ \hline 3 \end{array}$$

STEP 2 If you have enough of a small unit to change it to a larger unit, change it by dividing. Use the Table of Measurements to help you.

6 ft
+ 2 ft 3 in.
8 ft 3 in.

STEP 3 Combine your results.

Add the following measurements.

1. 2 hr 38 min
+ 6 hr 47 min

2. 2 gal 3 qt
+ 5 gal 3 qt

3. 4 lb 12 oz
+ 7 lb 11 oz

4. 8 yd 2 ft
+ 6 yd 1 ft

5. 3 days 18 hr
+ 8 days 16 hr

6. 2 m 75 cm
+ 3 m 45 cm

7. 8 ft 4 in.
3 ft 11 in.
+ 2 ft 9 in.

8. 4 kg 800 g
6 kg 250 g
+ 3 kg 415 g

9. 10 min 15 sec
5 min 52 sec
+ 9 min 47 sec

10. 3 T 1,500 lb
4 T 850 lb
+ 5 T 1,175 lb

11. 5 wk 4 days
6 wk 6 days
+ 4 wk 5 days

12. 1 pt 10 oz
3 pt 15 oz
+ 4 pt 12 oz

Subtracting Measurements

EXAMPLE   7 hr 35 min   **STEP 1** You can't take a larger number from a smaller
          − 4 hr 50 min   number, so you must borrow. Use the Table of
                                    Measurements to find out how many smaller units
                                    there are in the larger unit from which you must
                                    borrow. Borrow one larger unit changed to smaller
                                    units (1 hr = 60 min) and add it to the smaller units
                                    you already have (60 + 35 = 95 min).

          ^{6}7 hr $\overline{35}$ ^{95} min   **STEP 2** Subtract each column.
          − 4 hr 50 min
          ──────────────
          2 hr 45 min

Subtract the following measurements.

1.   8 ft 5 in.
    − 4 ft 8 in.

2.   5 days 11 hr
    − 3 days 18 hr

3.   13 lb 8 oz
    − 10 lb 14 oz

4.   12 gal 1 qt
    − 9 gal 3 qt

5.   9 m 45 cm
    − 5 m 92 cm

6.   8 min 25 sec
    − 2 min 53 sec

7.   10 pt 2 oz
    − 9 pt 8 oz

8.   3 T   800 lb
    − 2 T 1,560 lb

9.   6 km 375 m
    − 3 km 680 m

10.   5 yd
     − 2 yd 2 ft

11.   9 m
     − 8 m 36 cm

12.   9 wk
     − 4 wk 2 days

Multiplying Measurements

EXAMPLE	4 lb 10 oz	STEP 1	Multiply each unit of measurement separately.

$$\begin{array}{r} 4 \text{ lb } 10 \text{ oz} \\ \times \quad 3 \\ \hline 12 \text{ lb } 30 \text{ oz} \end{array}$$

STEP 2 If you have enough of a small unit to change it to a larger unit, change it by dividing. Use the Table of Measurements to help you.

$$\begin{array}{r} 1 \text{ lb } 14 \text{ oz} \\ 16 \overline{)30} \\ \underline{16} \\ 14 \end{array}$$

STEP 3 Combine your results.

$$\begin{array}{r} 12 \text{ lb} \\ + \ 1 \text{ lb } 14 \text{ oz} \\ \hline 13 \text{ lb } 14 \text{ oz} \end{array}$$

Multiply the following measurements.

1. $\begin{array}{r} 2 \text{ ft } 8 \text{ in.} \\ \times \quad 4 \\ \hline \end{array}$

2. $\begin{array}{r} 2 \text{ gal } 5 \text{ qt} \\ \times \quad 6 \\ \hline \end{array}$

3. $\begin{array}{r} 8 \text{ lb } 12 \text{ oz} \\ \times \quad 5 \\ \hline \end{array}$

4. $\begin{array}{r} 6 \text{ T } 350 \text{ lb} \\ \times \quad 10 \\ \hline \end{array}$

5. $\begin{array}{r} 8 \text{ m } 36 \text{ cm} \\ \times \quad 4 \\ \hline \end{array}$

6. $\begin{array}{r} 4 \text{ wk } 3 \text{ days} \\ \times \quad 6 \\ \hline \end{array}$

7. $\begin{array}{r} 2 \text{ yd } 8 \text{ in.} \\ \times \quad 9 \\ \hline \end{array}$

8. $\begin{array}{r} 3 \text{ km } 600 \text{ m} \\ \times \quad 12 \\ \hline \end{array}$

9. $\begin{array}{r} 7 \text{ pt } 11 \text{ oz} \\ \times \quad 3 \\ \hline \end{array}$

10. $\begin{array}{r} 4 \text{ kg } 640 \text{ g} \\ \times \quad 8 \\ \hline \end{array}$

11. $\begin{array}{r} 18 \text{ yd } 2 \text{ ft} \\ \times \quad 7 \\ \hline \end{array}$

12. $\begin{array}{r} 5 \text{ days } 18 \text{ hr} \\ \times \quad 6 \\ \hline \end{array}$

Dividing Measurements

EXAMPLE

$$
\begin{array}{r}
1 \text{ day} \quad\quad 13 \text{ hours} \\
4\overline{)6 \text{ days} \quad\quad 4 \text{ hours}} \\
\underline{4 \text{ days}} \\
2 \text{ days } = 48 \text{ hours} \\
\underline{+4} \\
52 \text{ hours} \\
\underline{52} \\
0
\end{array}
$$

STEP 1 Divide into the first unit.
$6 \div 4 = 1$ with a remainder of 2.

STEP 2 Change the remainder into the next type of units.

STEP 3 Add to the units you already have.

STEP 4 Divide again $52 \div 4 = 13$.

Divide the following measurements.

1. $6\overline{)8 \text{ lb } 4 \text{ oz}}$

2. $4\overline{)9 \text{ yd } 1 \text{ ft}}$

3. $5\overline{)17 \text{ gal } 2 \text{ qt}}$

4. $9\overline{)12 \text{ T } 300 \text{ lb}}$

5. $6\overline{)13 \text{ km } 800 \text{ m}}$

6. $4\overline{)15 \text{ ft } 8 \text{ in.}}$

7. $5\overline{)19 \text{ lb } 6 \text{ oz}}$

8. $12\overline{)30 \text{ wk } 6 \text{ days}}$

9. $13\overline{)28 \text{ m } 34 \text{ cm}}$

10. $17\overline{)36 \text{ min } 16 \text{ sec}}$

11. $4\overline{)9 \text{ kg } 420 \text{ g}}$

12. $11\overline{)24 \text{ pt } 1 \text{ oz}}$

Perimeter: Measuring the Distance Around a Rectangle

Most doors, windows, floors, and pieces of paper are in the shape of **rectangles.** A rectangle is a flat figure with four sides. The sides across from each other are equal in length and are also **parallel,** which means that they remain the same distance from each other. The long side of a rectangle is called the **length,** and the short side is called the **width.**

The distance around a rectangular shape is called the **perimeter.** To get the perimeter of a rectangular shape, add all four sides together.

EXAMPLE 1 Find the perimeter of the figure below.

length = 13 in.

width = 10 in.

> **STEP 1** Find out how long each side is. Since the sides across from each other are equal in length, you have two 13-inch sides and two 10-inch sides.

> **STEP 2** Add the lengths of all four sides.
> 13 + 13 + 10 + 10 = 46-inch perimeter

Another way to find the distance around a rectangle is to double the length and double the width. Then add these two figures together.

For the example above

2 × 10 in. = 20 in.
2 × 13 in. = 26 in.
20 in. + 26 in. = 46-inch perimeter

EXAMPLE 2 The distance around the outside of a room measures 40 feet. The width is 9 feet. How long is the room?

width = 9 ft   perimeter = 40 ft

In this problem, you already know the perimeter; that is, you know the total distance around the rectangle. You also know the distance covered by two of the four sides. You need to find the distance covered by one of the two remaining sides.

> **STEP 1** The distance covered by the two widths is 9 ft + 9 ft = 18 ft.

> **STEP 2** The distance covered by the two lengths is 40 ft – 18 ft = 22 ft.

> **STEP 3** The distance covered by one length is 22 ft ÷ 2 = 11 ft.

Solve and write the correct label, such as $ or feet, next to each answer.

1. How much fencing is required to enclose a garden that is 18 feet wide and 23 feet long?

2. Wire fencing costs $1.29 per foot. Find the cost of the fencing for the garden in the problem above.

3. Mr. Wiley wants to put weather stripping around the large windows of his house. Each window is 3 feet wide and 5 feet high. There are 11 of these windows in his house. How many feet of weather stripping must Mr. Wiley buy?

4. If the weather stripping costs 39¢ per foot, how much will it cost to buy weather stripping for the windows of Mr. Wiley's house?

5. Mr. Ellis wants to fence in part of his backyard for his young children. He has 80 feet of fencing to use. If the space he encloses is 17 feet wide, how long can it be?

6. How many inches of molding are needed to make a frame for a picture that is 24 inches wide and 37 inches long?

7. Mr. Romney's rectangular garden required 82 feet of fencing to completely enclose it. If the garden is 27 feet long, how wide is it?

Area: Measuring the Space Inside a Rectangle

When you want to decide how much carpeting to buy to cover your floor, or how much paint you need to cover a wall, you need to find **area.** Area is the measure of the amount of space inside a flat figure.

Area is measured in **square units:** square inches, square feet, square yards, square miles, and so on.

To find the area of something with a rectangular shape, multiply the length by the width.

EXAMPLE 1 Find the area of the rectangle shown below.

length = 5 in.

width = 3 in.

Area = 5 in. × 3 in. = 15 square inches

By dividing the rectangle in Example 1 into one-inch squares, you can see that there are 15 one-inch squares inside the rectangle.

To find the length or width of a rectangle when the area and either the length or width are given, *divide* the area by the side of the rectangle you know.

EXAMPLE 2 It took 240 square feet of carpeting to cover a room. If you know that the room is 20 feet long, how wide is it?

length = 20 ft

width = ?

Area = 240 sq ft

$$
\begin{array}{r}
12 \\
20\overline{)240} \\
\underline{20} \\
40 \\
\underline{40}
\end{array}
$$

 Solve and write the correct label, such as $ or square feet, next to each answer.

1. What is the area of a room that is 12 feet wide and 18 feet long?

2. Find the area in square yards for the room in problem 1. (1 square yard = 9 square feet)

3. If a carpet costs $16.99 per square yard, how much would it cost to buy carpet for the room in problem 1?

4. A store charges $45 to deliver and install carpet. Find the total cost including delivery and installation for the carpet in the last problem?

5. Mr. Cortez wants to put tiles on his basement floor. Each tile is 1 square foot in area. His basement floor is 63 feet long and 24 feet wide. How many tiles does he need to cover the basement floor?

6. If the tiles Mr. Cortez wants to use cost $6.75 per dozen, how much will the tiles for his basement floor cost?

7. Mr. Cortez needs 2 gallons of glue at $17.95 per gallon to attach the tiles. Find the combined cost of the tiles and the glue.

8. A group of neighbors plans to paint a wall that surrounds part of the playground of their community center. One gallon of paint is supposed to cover 300 square feet. If the wall is 12 feet high, what is the length of a section of the wall that can be covered by 1 gallon of paint?

9. The wall that the group wants to paint is 175 feet long. How many gallons of paint are needed to cover the wall?

10. The paint for the wall in the last two problems costs $19.95 per gallon. Find the total cost of the paint needed to cover the wall.

11. The area of the gym floor inside the community center is 7,168 square feet. If the gym floor is 64 feet wide, how long is it?

12. Find the area of the fenced-in play area in Mr. Ellis's yard in problem 5 on page 136.

13. Find the area of the picture in problem 6 on page 136.

14. What is the area of Mr. Romney's garden in problem 7 on page 136?

Volume: Measuring the Space Inside a 3-Dimensional Object

Things like boxes, containers, suitcases, and rooms have three dimensions. That means that, in addition to length and width, they also have height or depth. In order to find out the amount of space inside a 3-dimensional object, you have to find **volume.**

Volume is measured in **cubic units:** cubic inches, cubic feet, cubic yards, and so on.

To find the volume of a 3-dimensional object, such as a box or a trunk, multiply the length by the width by the height.

EXAMPLE

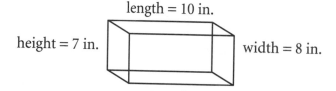

length = 10 in.

height = 7 in.

width = 8 in.

STEP 1 Multiply the length by the width.

$$\begin{array}{r} 10 \\ \times\ 8 \\ \hline 80 \end{array}$$

STEP 2 Multiply this figure by the height.

$$\begin{array}{r} 80 \\ \times\ 7 \\ \hline 560 \text{ cubic inches (cu in.)} \end{array}$$

 Solve and write the correct label, such as $ or cubic inches, next to each answer.

1. Mr. Miller is adding a room to the back of his house. For the foundation, he must dig a hole that is 16 feet long, 12 feet wide, and 4 feet deep. How many cubic feet of soil have to be removed?

2. Find the volume of a shoe box that is 9 inches wide, 15 inches long, and 6 inches high.

3. Mrs. Rodriguez's ice tray is 1 inch deep, 9 inches long, and 3 inches wide. How many cubic inches of ice will the tray hold?

4. Mrs. Rodriguez's tray makes ice cubes that are each 1 cubic inch. How many cubes can be made with five trays?

5. When the Corona family goes on vacation, they always pack one big trunk instead of a lot of small suitcases. Their trunk measures 6 feet long by 4 feet wide by 3 feet high. Find the volume of the packing space of the trunk.

6. The storeroom of the Apex Company is 30 feet long, 24 feet wide, and has a 12-foot high ceiling. How many cubic feet of storage space does the room contain?

7. How many cartons each measuring 3 feet by 3 feet by 2 feet can fit in the Apex Company storeroom?

8. To fill a hole in a vacant lot that is 12 feet long, 10 feet wide, and 10 feet deep, the community council hired a trucking company that could carry 40 cubic feet of dirt per truckload. How many truckloads will it take to fill the hole?

9. The trucking company charges $25 for the delivery of each truckload of dirt. What will be the cost of filling the hole in the last problem?

10. The book storage room in a training center is 16 feet long, 12 feet wide, and 8 feet high. How many boxes each measuring 4 feet by 4 feet by 2 feet can be put in the storage room?

Pricing a Meal from a Menu

Use the menu below to answer the questions on this page.

Lois's Restaurant Menu	
Soup of the day $2.75	Coffee $0.90
Tossed salad $3.25	Tea $0.80
Fried chicken dinner $7.95	Milk $0.95
Pork chop dinner $8.65	Pie $1.90
Steak dinner $9.75	Cake $1.65

1. When Fran went to lunch at Lois's, she ordered soup, a salad, and a cup of tea. How much did her lunch cost?

2. Carl took his afternoon break at Lois's. He had a piece of pie and a cup of coffee. How much was he charged?

3. Mr. Rigby took his secretary Mary out to lunch at Lois's. Mary had the pork chop dinner and a cup of coffee. Mr. Rigby had the steak, a cup of coffee, and a piece of cake. What was the total cost for Mary's and Mr. Rigby's lunches?

4. David had a big appetite when he went to Lois's. He ordered soup, a salad, the fried chicken dinner, a piece of pie, and two glasses of milk. How much did his lunch cost?

5. When Susan and Jim went to Lois's, they each had the fried chicken dinners. Susan also had a salad and a cup of tea. Jim had a cup of coffee and a piece of cake. Together, how much did their lunches cost?

Reading Sales Tax Tables

In many states and in some cities, businesses have to charge a tax on anything they sell. This tax is called a **sales tax.** To figure out how much sales tax to charge, businesses are given sales tax tables by the state or local government. The table below is a sample section of a sales tax table that is similar to the ones provided by the government. Although your state sales tax may be higher or lower than the tax shown on this table, every table is set up like the one shown below. In one column you must find the price of the item or items sold. Printed next to the total cost of the sale is the amount of sales tax that must be added. This table shows the sales tax on amounts from 1¢ to $5.91.

Amount	Tax	Amount	Tax	Amount	Tax
$0.01 to 0.08 -	0¢	$0.92 to 1.08 -	6¢	$1.92 to 2.08 -	12¢
0.09 to 0.24 -	1¢	1.09 to 1.24 -	7¢	2.09 to 2.24 -	13¢
0.25 to 0.41 -	2¢	1.25 to 1.41 -	8¢	2.25 to 2.41 -	14¢
0.42 to 0.58 -	3¢	1.42 to 1.58 -	9¢	2.42 to 2.58 -	15¢
0.59 to 0.74 -	4¢	1.59 to 1.74 -	10¢	2.59 to 2.74 -	16¢
0.75 to 0.91 -	5¢	1.75 to 1.91 -	11¢	2.75 to 2.91 -	17¢
$2.92 to 3.08 -	18¢	$3.92 to 4.08 -	24¢	$4.92 to 5.08 -	30¢
3.09 to 3.24 -	19¢	4.09 to 4.24 -	25¢	5.09 to 5.24 -	31¢
3.25 to 3.41 -	20¢	4.25 to 4.41 -	26¢	5.25 to 5.41 -	32¢
3.42 to 3.58 -	21¢	4.42 to 4.58 -	27¢	5.42 to 5.58 -	33¢
3.59 to 3.74 -	22¢	4.59 to 4.74 -	28¢	5.59 to 5.74 -	34¢
3.75 to 3.91 -	23¢	4.75 to 4.91 -	29¢	5.75 to 5.91 -	35¢

Suppose that you are a cashier in a restaurant or a store and you want to find the tax due on a $3.30 purchase. According to the table, the tax on an amount from $3.25 to $3.41 is 20¢; so the tax on $3.30 is 20¢.

Use the table to answer the following questions.

1. How much sales tax is due on each of the following amounts?

   a. $1.45

   b. $0.80

   c. $3.60

   d. $4.10

   e. $5.40

   f. $2.10

   g. $3.19

   h. $4.32

   i. $5.06

   j. $5.89

To find the sales tax for an amount larger than $5.91, first multiply the dollar amount by $0.06. Then add the tax for the remaining amount from the first section of the table on page 143.

<u>EXAMPLE</u>   Find the tax on $7.54.

$7 \times \$0.06 = \$0.42$ **STEP 1**   Multiply the dollar amount, 7, by $0.06.

$\quad\quad\quad\dfrac{+\,0.03}{\$0.45}$ **STEP 2**   Look up $0.54 on the tax table. The tax on $0.54 is 3¢. Add $0.42 and $0.03.

2. Find the tax on Fran's lunch (problem 1, page 142).

3. What was Fran's total lunch bill including tax?

4. How much change should Fran get back from $10?

5. Find the tax on Carl's pie and coffee (problem 2, page 142).

6. How much tax did Mr. Rigby have to pay on the bill for his and his secretary's lunches (problem 3, page 142)?

7. Mr. Rigby left a $3 tip. How much did he pay altogether for his and his secretary's lunches including tax and tip?

8. Find the tax on David's lunch (problem 4, page 142).

9. How much change should David receive from $20?

10. Find the tax on the lunch bill for Susan and Jim (problem 5, page 142).

11. Jim paid for his and Susan's lunches. How much change did he get back from $30?

Checking Your Change with Sales Receipts

On this page and the next are copies of sales receipts. On each receipt, a number followed by + indicates the price of an item. A number followed by TX means tax. A number followed by TL means a total.

 Use the sales receipts to answer the following questions. People who like to get rid of pennies should pay particular attention to the second problem for each receipt.

```
    Frank's Foods
    2930 Broadway

    $ 01.79   +
    $ 00.56   +
    $ 00.47   +
    $ 01.32   +

    $ 04.14   TL
    $ 00.33   TX

    $ 04.47   TL
```

1. How much change should you have received from $5?

2. How much change would you have received if you had given the clerk a $5 bill and 7¢ in change?

```
    Doug's Drugs
    4 East 110th

    $ 02.69   +
    $ 03.56   +
    $ 00.98   +

    $07.23   TL
    $00.58   TX

    $07.81   TL
```

3. How much change should you have received from $10?

4. If you paid with a $5 bill, three $1 bills, and a penny, how much change would you have gotten back?

```
    Bonnie's Books
    854 West George

    $ 01.95   +
    $ 03.50   +

    $ 05.45   TL
    $ 00.44   TX

    $ 05.89   TL
```

5. How much change should you have received from $10?

6. If you paid with a $5 bill, a $1 bill, and 4 pennies, how much change would you have received?

Hal's Hardware 5 East Third St.	
$09.59	+
$06.44	+
$08.27	+
$24.30	TL
$01.94	TX
$26.24	TL

7. How much change should you have gotten back from $30?

8. If you paid with a $20 bill, two $5 bills, two dimes, and four pennies, how much change would you have received?

Paula's Pastry 844 W. Oakdale	
$01.05	+
$02.15	+
$00.95	+
$04.15	TL
$00.33	TX
$04.48	TL

9. How much change should you have gotten back from $5?

10. If you paid with a $5 bill, one quarter, two dimes, and three pennies, how much change would you have gotten back?

Colette's Clothes 333 East 49th St.	
$07.29	+
$01.65	+
$13.99	+
$16.50	+
$39.43	TL
$03.15	TX
$42.58	TL

11. How much change should you have gotten back from $50?

12. If you paid with two $20 bills, a $5 bill, two quarters, a nickel, and three pennies, how much change would you have gotten back?

Using a Calorie Chart

Everything we eat has a certain number of calories. People who want to lose weight often have to cut down on the number of calories they take into their bodies each day. A doctor can tell you how many calories you need to stay healthy and lose weight at the same time. Dieters generally use a calorie chart to keep track of how many calories they are taking in.

 Below is a short calorie chart. Use this chart to answer the following questions.

Item	Quantity	Calories
chili con carne	1 small bowl	250
jelly doughnut	1	250
frankfurter	1	125
frankfurter roll	1	125
hamburger, fried	2-ounce patty	225
hamburger roll	1	250
milk shake	1	350
pretzels	5 small sticks	20
skim milk	8-ounce glass	90
cola	6 ounces	75
ginger ale	6 ounces	75
coffee with cream and sugar	1 cup	80

1. Martha stopped at the Doughnut Barn for a snack. She had two jelly doughnuts and coffee with cream and sugar. How many calories did she take in?

2. For her breakfast Marissa took in 600 calories, and for lunch, 715 calories. If she sticks to a limit of 2,400 calories per day, how many calories is she allowed for dinner?

3. Jeff stopped at the Come On Inn for breakfast. He had three jelly doughnuts, two 8-ounce glasses of skim milk, and one coffee with cream and sugar. How many calories did Jeff take in?

4. Before going home, Jeff went to Herb's Hamburgers where he ate two hamburgers on hamburger rolls and drank a milk shake. How many calories did Jeff take in at Herb's?

5. For lunch Alice had a small bowl of chili and a cola. How many calories did she take in for lunch?

6. Debbie's breakfast contained 280 calories. If she had a snack of 235 calories before dinner, and she is allowing herself only 2,000 calories a day, how many calories can Debbie take in at dinner?

7. At a baseball game, John had two frankfurters on rolls, a box of 50 pretzels, and four colas. How many calories did John take in at the game?

8. Helen had a breakfast of 475 calories. For lunch she had two hamburgers without rolls and a glass of ginger ale. If she limits herself to 2,000 calories a day, how many calories can there be in her dinner?

9. During his morning break, Felix had two jelly doughnuts and a cup of coffee with cream and sugar. During his afternoon break, he had a milk shake. How many calories did Felix take in during his breaks?

10. From noon to 6:00 P.M., Joel had a bowl of chili, two colas, a frankfurter on a roll, and a cup of coffee with cream and sugar. How many calories did he take in during this time?

Finding an Average

Finding an **average,** or **mean,** gives you information about a group of numbers.

EXAMPLE   Over a 4-week period, a waiter worked 40 hours one week, 25 hours the next week, 45 hours the week after, and 30 hours during the fourth week. To figure out how many hours it evens out to per week, you can find an average.

To find the average of a group of numbers, add the numbers together and divide by the number of numbers you added. Find the average number of hours the waiter worked per week.

STEP 1   Add the numbers you want to average.

$$\begin{array}{r} 40 \\ 25 \\ 45 \\ + 30 \\ \hline 140 \end{array}$$

STEP 2   Divide by the number of numbers you added.

$$\overset{\displaystyle 35 \text{ hours per week}}{4\overline{)140}}$$

ANSWER: On an average, the waiter worked **35 hours per week** over the 4-week period.

Solve and write the correct label, such as $ or hours, next to each answer.

1. At night, Pete drives a cab. Monday night he drove for 3 hours, Tuesday night for 6 hours, Wednesday night for 5 hours, Thursday night for 4 hours, and Friday night for 7 hours. What was the average number of hours he drove each night?

2. On Monday night, Pete made $14.30 in tips; on Tuesday, $28.55; on Wednesday, $26.15; on Thursday, $21.65; and on Friday, $42.35. How much did Pete average in tips per night?

3. During a basketball tournament, the number of tickets sold each night was: first night—4,065 tickets; second night—3,983 tickets; third night—4,117 tickets; and the last night—5,267 tickets. What was the average number of tickets sold for each night of the tournament?

4. Elizabeth received the following scores on math tests: 86, 76, 93, 89, 68, and 92. What was the average of her math test scores?

5. The tenants' association of the Kendal Projects meets every Tuesday evening. On the 4th there were 46 people at the meeting. On the 11th there were 53 people. On the 18th there were 38 people. And on the 25th there were 55 people. What was the average attendance for meetings that month?

 6. In 1996 Mr. Lee made $26,776; in 1997 he made $29,231; in 1998 he made $18,385; and in 1999 he made $27,595. Round each year's income to the nearest hundred dollars. Then estimate his average annual income for these years.

7. In 1996 Mr. Lee's family consisted of his wife, their two children, and himself. Using the exact income, find the average (per capita) income for each member of the Lee family in 1996.

8. In 1999 the Lees had another child. Using the exact income, find the per capita income for each member of the Lee family in 1999.

9. The Perez family took 3 days to drive to their grandparents' house in the Midwest. Thursday they drove 487 miles, Friday they drove 392 miles, and Saturday they drove 456 miles. What was the average distance that they drove each day?

10. On Thursday the Perez family spent $136 for gas, food, and lodging. They spent $157 on Friday, and $52 on Saturday. Find their average daily expenses.

Reading Paycheck Stubs

Below is a copy of a weekly paycheck stub from Addie Strand's paycheck. It shows her gross pay (before deductions) and her net pay (after deductions). It also shows the amounts withheld for FICA (social security) and various taxes for both the pay period shown and the year to date.

Use this check stub to answer the following questions.

Edison	STATEMENT OF EARNINGS AND DEDUCTIONS					DETACH AND RETAIN FOR YOUR RECORDS	

CHECK NUMBER 0032160 | ROLL 93077 | LOC 708X

PERIOD ENDING	EMPLOYEE NAME	EMPLOYEE NUMBER
MO DAY YEAR		
05 28 99	A STRAND	40622

TAXES WITHHELD					GROSS PAY		GROSS PAY TOTAL
	FEDERAL	FICA	STATE	CITY	TAXABLE FOR F.I.C.A	NON TAXABLE FOR F.I.C.A	
CURRENT	89.70	24.91	19.94	9.96	498.34		498.34
	FEDERAL	FICA	STATE	CITY	TAXABLE FOR F.I.C.A	YEAR TO DATE EARNINGS	NET PAY
YEAR TO DATE	1973.40	548.00	438.70	219.10	10963.48	10963.48	
							353.83

1. What is Ms. Strand's gross pay each week?

2. What is Ms. Strand's net pay each week?

3. How much has been withheld to date for federal tax?

4. How much has been withheld to date for social security?

5. How much has been withheld to date for state tax?

6. What is Ms. Strand's gross pay to date?

7. If Ms. Strand makes the same amount for 52 weeks in the year, what is her yearly gross pay?

8. If she has the same deductions all year, what is Ms. Strand's net pay for the year?

9. There are 30 more pay periods for Ms. Strand for the year indicated on the paycheck stub. How much take-home pay will Ms. Strand receive for the rest of the year?

10. The same amount is withheld each week from her check. Find the total amount of federal tax that is withheld from Ms. Strand's pay in one year.

11. Ms. Strand budgets $165 a week for food for her children and herself. How much is left from her take-home pay each week after paying for food?

12. Ms. Strand's house mortgage amounts to $107.50 per week. How much does she have left from her weekly net pay after paying for food and the mortgage?

13. Ms. Strand's car payments are $38.60 per week. How much does she have left from her weekly net pay after paying for food, the mortgage, and the car?

14. Which of the following approximates the amount Ms. Strand has left each *month* for clothes, utilities, savings, entertainment, and miscellaneous expenses?

    a. $345
    b. $295
    c. $215
    d. $175

Using a Table to Look Up Check Cashing Rates

If you cash a check at a private check cashing establishment rather than at a bank or at the company where you work, you will be charged a fee. This fee is deducted from the amount of the check.

The table below shows the fees to be charged at a certain check cashing establishment for checks from $1.00 to $204.99.

Honest John's Check Cashing		
$1.00 to 14.99 – 20¢	$ 75.00 to 79.99 – $0.85	$140.00 to 144.99 – $1.50
15.00 to 19.99 – 25¢	80.00 to 84.99 – 0.90	145.00 to 149.99 – 1.55
20.00 to 24.99 – 30¢	85.00 to 89.99 – 0.95	150.00 to 154.99 – 1.60
25.00 to 29.99 – 35¢	90.00 to 94.99 – 1.00	155.00 to 159.99 – 1.65
30.00 to 34.99 – 40¢	95.00 to 99.99 – 1.05	160.00 to 164.99 – 1.70
35.00 to 39.99 – 45¢	100.00 to 104.99 – 1.10	165.00 to 169.99 – 1.75
40.00 to 44.99 – 50¢	105.00 to 109.99 – 1.15	170.00 to 174.99 – 1.80
45.00 to 49.99 – 55¢	110.00 to 114.99 – 1.20	175.00 to 179.99 – 1.85
50.00 to 54.99 – 60¢	115.00 to 119.99 – 1.25	180.00 to 184.99 – 1.90
55.00 to 59.99 – 65¢	120.00 to 124.99 – 1.30	185.00 to 189.99 – 1.95
60.00 to 64.99 – 70¢	125.00 to 129.99 – 1.35	190.00 to 194.99 – 2.00
65.00 to 69.99 – 75¢	130.00 to 134.99 – 1.40	195.00 to 199.99 – 2.05
70.00 to 74.99 – 80¢	135.00 to 139.99 – 1.45	200.00 to 204.99 – 2.10

Find the fee for cashing checks for each of the following amounts.

1. $43.89

2. $80.10

3. $98.50

4. $153.99

5. $104.85

6. $203.57

7. $76.99

8. $119.60

9. $134.95

10. $187.92

11. Hector took his paycheck of $137.50 to Honest John's. How much cash should Hector receive?

12. Sonia cashed her check for $193.65 at Honest John's. She also paid her gas and electricity bill for $42.47. How much did she take home from Honest John's?

13. Doreen cashed her check for $142.75 and paid her phone bill for $36.58 at Honest John's. How much cash did she take home?

14. Tony cashed his check for $198.70 and his wife's check for $168.45 at Honest John's. If he had to pay a separate charge for each check, how much did Tony take home for himself and his wife?

15. Danny cashed his $184.00 tax refund and paid his $34.65 phone bill and $29.42 gas and electricity bill at Honest John's. How much did he take home?

16. Frank cashed his check from his part-time job for $174.29 at Honest John's. Frank's brother Pete went with him to collect the $50.00 Frank had borrowed a month before. How much did Frank take home after paying the money he owed Pete?

17. Andrea cashed her paycheck for $182.60 and her rebate of $25.00 for her new TV at Honest John's. She had to pay a separate fee for each check. How much money did she take home?

Understanding Tax Statements

At the beginning of every year, employers give their employees a wage and tax statement (W-2 Form), which must be attached to everyone's federal and state income tax forms. The wage and tax statement gives the following information:

1. the total amount of the employee's salary before deductions were taken out, called **gross salary** (shown in box 2)
2. the amount deducted for federal income tax (shown in box 1)
3. the amount deducted for FICA, or social security (shown in box 3)
4. the amount deducted for state income tax (shown in box 6)
5. the amount deducted for city income tax (shown in box 9)

The year shown in the upper right-hand corner of the statement is the year in which the salary and taxes were paid.

The wage and tax statement below shows how much Adam Smith earned in 1999 while he was working at Positive Systems in New York.

Use this statement to answer the following questions.

POSITIVE SYSTEMS
150 NASSAU ST.
NEW YORK, N.Y.

Wage and Tax Statement **1999**

Copy C   For
employee's records

Type or print EMPLOYER'S Federal identifying number, name, address and ZIP code above.

FEDERAL INCOME TAX INFORMATION				SOCIAL SECURITY INFORMATION				STATE OR LOCAL TAX INFORMATION					
1	Federal income tax withheld	2	Wages, tips and other compensation	3	FICA employee tax withheld	4	Total FICA wages	6	State Tax withheld	7	Wages paid	8	Name of State
	5,271.32		29,285.00		1,698.53		29,285.00		1,171.42				NEW YORK

Type or print EMPLOYEE'S social security no., name, and address including ZIP code below

123–45–6789

SMITH, ADAM
950 W. 57th ST.
N.Y., N.Y. 10019

5	Uncollected employee FICA tax on tips	9	City tax withheld	10	Wages paid	11	Name of City
			409.99				NEW YORK

Employee No.	Form No.	Was employee covered by a qualified pension plan etc.?
	State IT-2102	CITY OR OTHER NYC–2	Yes ☒ No ☐

OTHER INFORMATION (SEE CIRCULAR E)			
Cost of group term life insurance included in box 2.7.10	Excludable sick pay included in box 2.7.10		Contribution to individual employee retirement account

Form **W-2**   An "X" in the upper left corner indicates this is a corrected form.
22 1766474

This information is being furnished to the Internal Revenue Service and appropriate State officials.

1. How much did Mr. Smith make during 1999 before deductions?

2. How much was deducted from Mr. Smith's wages for FICA (social security)?

3. How much was withheld from Mr. Smith's wages for federal income tax?

4. How much was withheld from Mr. Smith's wages for state income tax?

5. How much was withheld from Mr. Smith's wages for city income tax?

6. What were the total deductions taken from Mr. Smith's wages in 1999?

7. What was Mr. Smith's net income (take-home pay) for 1999?

8. To the nearest ten dollars, what was Mr. Smith's average monthly take-home pay?

9. Mr. Smith and his wife pay $572.60 a month for rent. How much do they spend on rent in one year?

10. How much is left from Mr. Smith's yearly take-home income after rent is paid?

11. The average food bill for Mr. and Mrs. Smith is $118 per week. How much do they spend on food in one year? (1 year = 52 weeks)

12. How much is left from Mr. Smith's take-home pay when both rent and food are deducted?

13. The average monthly gas and electricity bills for the Smiths is $41.50. How much do they pay for gas and electricity in one year?

14. How much is left from Mr. Smith's take-home pay after rent, food, gas, and electricity are deducted?

Figuring the Cost of Electricity

The chart below shows the average costs for operating various electrical appliances. Charts like this one are distributed by power companies to their customers, although rates are different in different cities. Using a chart like this helps people to understand and keep down their electricity bills.

Use this chart to answer the following questions.
(Use 1 month = 30 days.)

Appliance	Average Cost	Appliance	Average Cost
Refrigerator (single door)		Blender	47 uses for 1¢
12 cu ft, manual defrost	18¢ a day	Mixer (hand)	18 uses for 1¢
Refrigerator/Freezer (two door)		Television	3¢ per hour
		Lightbulb	
14 cu ft, cycle defrost	31¢ a day	100-watt	1.2 hours for 1¢
14 cu ft, frostless	37¢ a day	60-watt	2 hours for 1¢
18 cu ft, frostless	47¢ a day	25-watt	4.7 hours for 1¢
Freezer		Room air conditioner	13¢ per hour
15 cu ft chest,		Electric blanket	9¢ per night
manual defrost	34¢ a day	Clock	15¢ per month
16 cu ft upright,		Vacuum cleaner	5¢ per hour
manual defrost	44¢ a day	Washing machine	2¢ per load
16 cu ft upright,		Dryer	26¢ per load
frostless	60¢ a day	Iron	5¢ per hour
Dishwasher	5¢ per use		
Toaster	4 slices for 1¢		

1. How much does it cost to run a television for 3 hours?

2. If Lynne watches her television an average of 3 hours every night, how much does it cost her to watch television in one month?

3. How much does it cost to keep a 12-cubic-foot single-door refrigerator running for one month?

4. How much does it cost to use an iron for 8 hours?

5. Find the cost of burning four 60-watt bulbs for 6 hours.

6. How much would it cost to burn four 60-watt light bulbs for 6 hours every day for a month?

7. During the summer months, the Rosas use their room air conditioner 9 hours a day. How much does it cost them to run their air conditioner for 9 hours?

8. If the Rosas use their air conditioner 9 hours every day for a month, how much will it cost them?

9. How much does it cost to toast 100 slices of bread?

10. Following is an account of how much electricity the Frey family used in one month. Figure out the cost for each appliance and the monthly electricity bill.

Appliance	Time Use	Monthly Cost
14 cu ft frostless refrig/freezer	all the time	_____
television	2 hours per day	_____
four 60-watt lightbulbs	6 hours per day	_____
electric clock	all the time	_____
iron	4 hours a month	_____
vacuum cleaner	10 hours a month	_____
	Total	_____

Reading Time Schedules

NEW YORK/BOSTON							
	Leave	Arrive	Flight No.	Stops or Via	Meals	Equip.	Freq.
To Boston	7:00a L	7:53a	308	NON-STOP	☕	727	ExSaSu
	7:15a K	8:09a	10	NON-STOP	☕	D10	Daily
F $173.00	8:35a L	9:32a	496	NON-STOP	☕	727	ExSu
Y $112.00	9:45a L	10:37a	360	NON-STOP		727	Daily
	11:50a L	12:42p	606	NON-STOP	🍷	727	Daily
	1:20p L	2:10p	82	NON-STOP	🍷	727	Daily
	4:05p L	4:59p	579	NON-STOP	🍷	727	Daily
	5:10p K	6:07p	186	NON-STOP	🍷	707	Daily
	5:40p K	6:43p	126	NON-STOP	🍷	D10	Daily
	7:30p L	8:24p	228	NON-STOP	🍷	727	Daily
	8:30p L	9:19p	276	NON-STOP	🍷	727	ExSa
	9:30p L	10:19p	598	NON-STOP	🍷	727	ExSa
	10:00p K	10:57p	118	NON-STOP	🍷	707	Daily
	1:10a K	2:01a	4	NON-STOP	🍷	707	ExSu
From Boston	7:00a	7:52a L	577	NON-STOP	☕	727	ExSaSu
	7:35a	8:29a K	117	NON-STOP	☕	707	Daily
	8:40a	9:32a L	599	NON-STOP	☕	727	ExSu
	10:35a	11:27a L	85	NON-STOP	🍷	727	Daily
	11:25a	12:17p L	369	NON-STOP	🍷	727	Daily
	1:20p	2:11p L	249	NON-STOP	🍷	727	Daily
	2:45p	3:41p K	187	NON-STOP	🍷	707	Daily
	4:05p	4:57p L	261	NON-STOP	🍷	727	ExSa
	5:15p	6:07p L	511	NON-STOP	🍷	727	Daily
	6:50p	7:43p L	341	NON-STOP	🍷	727	ExSa
	6:50p	7:43p L	329	NON-STOP	🍷	727	Sa
	8:25p	9:15p L	545	NON-STOP	🍷	727	ExSa
	9:00p	9:58p K	95	NON-STOP	🍷	707	Daily

The schedule above is for one airline's flights between the New York City airports and Boston. Following is a list of the meanings of the abbreviations used in the schedule:

F—first class ticket (one way)
Y—coach ticket (one way)
a—morning (A.M.)
p—afternoon/evening (P.M.)
K—Kennedy airport
L—LaGuardia airport

Flight No.—flight number
Equip.—type of airplane
Freq.—how often the flight leaves (what days)
Ex—except

Use the New York/Boston schedule to answer the following questions.

1. At what time does flight number 606 leave from New York?

2. At what time does flight number 606 arrive in Boston?

3. How long does flight number 606 take to get to Boston? (Remember to borrow 60 minutes.)

4. How long does flight 341 take to get from Boston to New York?

5. Does flight 228 go to Boston every day?

6. Does flight 577 go to New York every day?

7. How long does the earliest flight leaving from Kennedy airport to Boston take to get to Boston?

8. Mr. Parker paid $13.00 for a bus ride from midtown New York to LaGuardia Airport and $22.50 for a taxi ride from Boston's airport to downtown Boston. If he buys a coach ticket, how much does the trip from midtown New York to downtown Boston cost Mr. Parker?

9. If Mr. Parker pays the same fares for his return, what is the round-trip cost from midtown New York to downtown Boston and back to midtown New York?

10. Ms. Ramos took flight number 360 to Boston and returned the same day on flight number 511. How much time did she have in Boston? (Find how much time before noon and how much after noon.)

11. How much is round-trip coach fare between New York and Boston?

12. How much is round-trip first-class fare between New York and Boston?

13. How much does a passenger save on round-trip fare between New York and Boston by going coach rather than first class?

Renting a Car

Many car rental agencies charge different rates for different types of cars. Your car rental bill will also depend on which days of the week you rent the car as well as the number of days you keep it. Notice that Ralph's Rent-a-Car (in the ad shown below) offers unlimited free mileage. That means that there is no charge for the distance the car is driven.

 Use the ad below to answer the following questions.

	Ralph's Rent-a-Car Unlimited Free Mileage		
Model	Monday to Thursday (Per Day)*	Fri., Sat., Sun. & Holidays (Per Day)*	Weekly Any 7 Cons. Days
Compact	$15.95	$22.95	$110.95
Deluxe Compact	$18.95	$25.95	$125.95
Intermediate	$20.95	$27.95	$130.95
Standard	$22.95	$28.95	$145.95

*You will be charged the full day rate regardless of the time of day you pick up and return the car.

1. On Friday morning, Morris rented a deluxe compact from Ralph's Rent-a-Car. He returned it Sunday night. How much was he charged?

2. Mr. and Mrs. Colon rented a standard car from Ralph's. They picked up the car on Wednesday and returned it the next Tuesday. How much did they have to pay at Ralph's?

3. How much would it cost to rent an intermediate size car from Sunday night until noon on Thursday?

4. How much does it cost to rent a compact car for the 3-day Labor Day weekend?

5. Mr. Slater rented a compact car Friday night and returned it on Sunday. During that time, he filled the tank three times for the following prices: $8.50, $6.90, and $9.15. How much did he spend on gasoline and car rental for that weekend?

6. Mr. and Mrs. Sosa rented a standard size car from Ralph's on Sunday evening and returned it during the day on Thursday. If they bought 40 gallons of gas at $1.35 a gallon, how much did they pay for the car including gas and rental fee?

Using a Calculator

The goal of this book is to improve your mental math skills and your pencil and paper math skills. However, as we all know, a calculator is a useful tool for getting quick, exact answers.

EXAMPLE 1   Use a calculator to find the sum of $463 + 28 + 7$.

Press the following keys on a calculator: | 4 | 6 | 3 | + | 2 | 8 | + | 7 | = |

The calculator display should read | 498. |

EXAMPLE 2   Use a calculator to find the difference of $8,047 - 529$.

Press the following keys on the calculator: | 8 | 0 | 4 | 7 | − | 5 | 2 | 9 | = |

The calculator display should read | 7518. |

EXAMPLE 3   Use a calculator to find the product of 6×793.

Press the following keys on the calculator: | 6 | × | 7 | 9 | 3 | = |

The calculator display should read | 4758. |

EXAMPLE 4   Use a calculator to find the quotient of $13,244 \div 28$.

Press the following keys on the calculator: | 1 | 3 | 2 | 4 | 4 | ÷ | 2 | 8 | = |

The calculator display should read | 473. |

Calculators do not show remainders. For division problems that do not divide evenly, the screen shows a decimal answer.

EXAMPLE 5   Use a calculator to find the quotient of $428 \div 12$.

Press the following keys on the calculator: | 4 | 2 | 8 | ÷ | 1 | 2 | = |

The calculator display should read | 35.666666 |

Using Mental Math

Mental math refers to the problems you can do without pencil and paper and without a calculator. Below are examples of mental math problems that are covered in this book.

THE BASIC ARITHMETIC FACTS

These are the most important tools in mathematics. Memorize any facts that you do not know. Learning the arithmetic facts is the key to your success in studying more mathematics.

Addition	Subtraction	Multiplication	Division
$5 + 9 = 14$	$12 - 3 = 9$	$6 \times 9 = 54$	$42 \div 6 = 7$
$8 + 3 = 11$	$9 - 2 = 7$	$8 \times 7 = 56$	$18 \div 2 = 9$
$7 + 6 = 13$	$11 - 6 = 5$	$4 \times 3 = 12$	$54 \div 9 = 6$
$1 + 5 = 6$	$8 - 7 = 1$	$9 \times 2 = 18$	$24 \div 3 = 8$
$4 + 7 = 11$	$15 - 6 = 9$	$5 \times 12 = 60$	$40 \div 8 = 5$

COMPATIBLE PAIRS FOR ADDITION

When you have to add several numbers in one problem, look for **compatible pairs.** Compatible pairs are two numbers whose sum ends in zero. They are easy to add in your head.

$$24 + 6 = 30 \qquad 9 + 41 = 50 \qquad 82 + 8 = 90 \qquad 3 + 37 = 40$$

QUICK SUBTRACTION WITH ZEROS

Watch for subtraction problems in which the subtrahend (the second or bottom number) has a zero. These problems are easy to subtract in your head because you do not have to borrow. Remember that zero subtracted from any number is that number.

$$83 - 60 = 23 \qquad 92 - 50 = 42 \qquad 27 - 10 = 17 \qquad 56 - 20 = 36$$

MULTIPLYING BY 10, 100, AND 1,000

To multiply a number by 10, put one zero to the right of the number.

$$25 \times 10 = 250 \qquad 10 \times 40 = 400 \qquad 360 \times 10 = 3,600$$

To multiply a number by 100, put two zeros to the right of the number.

$$48 \times 100 = 4,800 \qquad 100 \times 136 = 13,600 \qquad 750 \times 100 = 75,000$$

To multiply a number by 1,000, put three zeros to the right of the number.

$$1,000 \times 9 = 9,000 \qquad 72 \times 1,000 = 72,000 \qquad 1,000 \times 300 = 300,000$$

QUICK MULTIPLICATION WITH ZEROS

To multiply numbers that end in zeros, first multiply the *non-zero* numbers. Then bring along the zeros from the original numbers.

$50 \times 7 = 350$ $40 \times 300 = 12{,}000$ $80 \times 120 = 9{,}600$

QUICK DIVISION WITH ZEROS

When both the dividend and the divisor in a division problem end in zeros, you can *cancel* the zeros one-for-one.

$$\frac{30\cancel{0}}{6\cancel{0}} = 5 \qquad \frac{2{,}70\cancel{0}}{30\cancel{0}} = 9 \qquad \frac{14{,}00\cancel{0}}{7\cancel{0}} = 200 \qquad \frac{4{,}80\cancel{0}}{12\cancel{0}} = 40$$

PROPERTIES OF ONE AND ZERO

Any number multiplied by one is that number.

$1 \times 9 = 9$ $8 \times 1 = 8$ $1 \times 36 = 36$ $12 \times 1 = 12$

Any number divided by one is that number.

$7 \div 1 = 7$ $5 \div 1 = 5$ $48 \div 1 = 48$ $13 \div 1 = 13$

Zero divided by any number is zero.

$0 \div 3 = 0$ $0 \div 52 = 0$ $0 \div 8 = 0$ $0 \div 25 = 0$

Any number divided by zero is said to be *undefined*.

$7 \div 0 = \text{undefined}$ $5 \div 0 = \text{undefined}$ $3 \div 0 = \text{undefined}$ $12 \div 0 = \text{undefined}$

Using Estimation

Estimating means finding an approximate answer. Estimating is a way to check whether an answer is reasonable. Most methods of estimating involve rounding the numbers in a problem to numbers that end in zeros.

To round a whole number,

STEP 1 Underline the digit in the place you are rounding to.

STEP 2 **a.** If the digit to the right of the underlined digit is *5 or more,* add 1 to the underlined digit.

b. If the digit to the right of the underlined digit is *less than 5,* leave the underlined digit as is.

STEP 3 Change all the digits to the right of the underlined digit to zeros.

ESTIMATING ADDITION PROBLEMS

One way to estimate an answer to an addition problem is to round each number to the same place.

EXAMPLE 1 Estimate the answer to the problem $3,143 + 571 + 927$ by rounding each number to the nearest hundred.

$3,143 + 571 + 927 \approx$ **STEP 1** Round each number to the
$3,100 + 600 + 900 =$ nearest hundred.
$\quad 4,600$ **STEP 2** Add the rounded numbers.

(The exact answer is 4,641.)

Another way to estimate is called **front-end rounding.** To get a quick estimate of an answer, round each number in a problem to the *left-most* place. Then do the operation with the rounded numbers.

EXAMPLE 2 Use front-end rounding to estimate the answer to the problem $52 + 873 + 2,184$.

$52 + 873 + 2,184 \approx$ **STEP 1** Round 52 to the nearest ten,
$50 + 900 + 2,000 =$ 873 to the nearest hundred,
  and 2,184 to the nearest
  thousand.
$\quad 2,950$ **STEP 2** Add the rounded numbers.

(The exact answer is 3,109.)

ESTIMATING SUBTRACTION PROBLEMS

To estimate the answer to a subtraction problem, round each number to the same place.

EXAMPLE 1 Estimate the answer to the problem 28,396 – 7,564 by rounding each number to the nearest thousand.

$$28,396 - 7,564 \approx$$
$$28,000 - 8,000 =$$
$$20,000$$

STEP 1 Round each number to the nearest thousand.

STEP 2 Subtract the rounded numbers.

(The exact answer is 20,832.)

EXAMPLE 2 Use front-end rounding to estimate the answer to the problem 4,318 – 879.

$$4,318 - 879 \approx$$
$$4,000 - 900 =$$
$$3,100$$

STEP 1 Round each number to the left-most digit.

STEP 2 Subtract the rounded numbers.

(The exact answer is 3,439.)

ESTIMATING MULTIPLICATION PROBLEMS

One way to estimate the answer to a multiplication problem is to round each number in the problem to the same place.

EXAMPLE 1 Estimate the answer to the problem 39×123 by rounding each number to the nearest ten.

$$39 \times 123 \approx$$
$$40 \times 120 =$$
$$4,800$$

STEP 1 Round each number to the nearest ten.

STEP 2 Multiply the rounded numbers.

(The exact answer is 4,797.)

EXAMPLE 2 Use front-end rounding to estimate the answer to the problem $57 \times 3,194$.

$$57 \times 3,194 \approx$$
$$60 \times 3,000 =$$
$$180,000$$

STEP 1 Round each number to the left-most digit.

STEP 2 Multiply the rounded numbers.

(The exact answer is 182,058.)

ESTIMATING DIVISION PROBLEMS

One way to estimate the answer to a division problem is to round the dividend (the number being divided) to a number that divides evenly by the divisor.

EXAMPLE 1 Estimate the answer to the problem $358 \div 9$.

$358 \div 9 \approx$ **STEP 1**   Round 358 to the nearest ten.
$360 \div 9 =$
    40   **STEP 2**   Divide 360 by 9.

(The exact answer is 39 r 7.)

However, rounding to the nearest ten, hundred, or thousand does not always give a number that divides evenly by the divisor.

EXAMPLE 2 Estimate the answer to the problem $491 \div 6$.

$491 \div 6 \approx$ **STEP 1**   491 to the nearest ten is 490, and 491 to the nearest hundred is
$480 \div 6 =$ 500. But 6 does not divide evenly into 490 or 500. The closest
                             round number that 6 divides into evenly is 480.

    80   **STEP 2**   Divide 480 by 6.

(The exact answer is 81 r 5.)

Another way to estimate the answer to a division problem is to do a partial division. Instead of completing a division problem, try for **two-digit accuracy.** Divide until you have three digits in the quotient. Then round your answer.

EXAMPLE 3 Estimate the answer to the problem $13,269 \div 8$.

$$\begin{array}{r} 1{,}65- \approx 1{,}700 \\ 8\overline{)13{,}269} \\ \underline{8} \\ 5\,2 \\ \underline{4\,8} \\ \overline{46} \end{array}$$

STEP 1   Ask yourself how many times 8 goes into 13. $13 \div 8 = 1$ with a remainder.

STEP 2   Place 1 over the 3 and multiply $1 \times 8 = 8$.

STEP 3   Subtract $13 - 8 = 5$.

STEP 4   Bring down the 2.

STEP 5   Ask yourself how many times 8 goes into 52. $52 \div 8 = 6$ with a remainder.

STEP 6   Place 6 over the 2 and multiply $6 \times 8 = 48$.

STEP 7   Subtract $52 - 48 = 4$.

STEP 8   Bring down the 6. Then decide whether 8 goes into 46 5 times or more. $46 \div 8 = 5$ with a remainder. Since the digit to the right of 6 is 5, raise 6 to 7, and put zeros in the tens and units places.

(The exact answer is 1,658 r 5.)

Units of Measurement

The table below lists common units of measure and what they are equal to in other units. The terms *meter, kilometer, kilogram,* and *liter* may not be familiar to you. They are part of the **metric system** of measurement. The metric system is used in many countries and frequently in the U.S. However, the **U.S. customary system** of measurement is the one more commonly used here.

Common abbreviations for each unit of measurement are given in parentheses.

Measures of Length

1 foot (ft or ')	= 12 inches (in. or ")
1 yard (yd)	= 36 inches
1 yard	= 3 feet
1 mile (mi)	= 5,280 feet
1 mile	= 1,760 yards
1 meter (m)	= 1,000 millimeters (mm)
1 meter	= 100 centimeters (cm)
1 kilometer (km)	= 1,000 meters

Measures of Weight

1 pound (lb)	= 16 ounces (oz)
1 ton (T)	= 2,000 pounds
1 kilogram (kg)	= 1,000 grams (gr)

Liquid Measures

1 pint (pt)	= 16 ounces
1 pint	= 2 cups
1 quart (qt)	= 2 pints
1 gallon (gal)	= 4 quarts
1 liter	= 1,000 milliliters

Measures of Time

1 minute (min)	= 60 seconds (sec)
1 hour (hr)	= 60 minutes
1 day	= 24 hours
1 week (wk)	= 7 days
1 year (yr)	= 365 days

Glossary

A

addend   One of the numbers in an addition problem. In 3 + 2 = 5 the addends are 3 and 2.

addition   The mathematical operation used to find a sum. The problem 8 + 1 = 9 is an example.

approximate   Another word for *estimate.* As an adjective it means "close or almost exact." A jacket costs $79. The approximate price is $80. The symbol ≈ means "approximately equal to."

area   A measure of the amount of surface on a flat figure. A room that is 12 feet long and 10 feet wide has an area of 12 × 10 = 120 square feet.

average   A sum divided by the number of items that make up the sum. An average, also called the *mean,* is a representative number for a group. If the low temperature one day was 24° and the low temperature the next day was 28°, the average low temperature for the two days is the sum (24° + 28° = 52°) divided by 2 (the number of days). 52° ÷ 2 = 26°

B

borrowing   Regrouping the digits in the top number of a subtraction problem. In the problem 84 − 29, the 8 in the tens column becomes 7, and the 4 in the units column becomes 14.

C

carrying   Regrouping the digits in an addition or multiplication problem. In the problem 37 + 9, the sum of the units columns is 7 + 9 = 16. The digit 6 remains in the units column, and the digit 1 is added to the tens column.

commutative property for addition   Numbers can be added in any order. The sum of 4 + 9 is the same as the sum of 9 + 4.

commutative property for multiplication   Numbers can be multiplied in any order. The product of 3 × 6 is the same as the product of 6 × 3.

compatible pairs   Two numbers with which it is easy to perform a mathematical operation. The numbers 83 and 7 are a compatible pair. They are easy to add because their sum ends with zero.

cubic units   A category of measurement for volume. The volume of a tank is usually measured in cubic inches or cubic feet.

D

difference   The answer to a subtraction problem. In the problem 14 − 9 = 5, the difference is 5.

digit   One of the ten number symbols. The digits are 0, 1, 2, 3, 4, 5, 6, 7, 8, and 9.

distance   A measure of the length of a straight line between two points. The distance between Chicago and Milwaukee is 90 miles.

dividend   The number in a division problem into which another number divides. In 21 ÷ 7 = 3, the dividend is 21.

division   A mathematical operation that requires figuring out how many times one amount is contained in another. In the problem 20 ÷ 5 = 4, the answer means that there are exactly four fives in twenty.

divisor   The number in a division problem that divides into another. In 28 ÷ 7 = 4, the divisor is 7.

E

estimate   As a noun: an approximate value. The population of a village is 1,836. The number 1,800 is an estimate of the population. As a verb: to find an approximate value. You may be asked to estimate the number of people living on your street.

F

factor   A number that, when multiplied by another number, results in a product. The numbers 2 and 3 are both factors of 6.

front-end rounding   Rounding the left-most digit of each number in a problem in order to calculate an estimate. In the problem 87×638, the number 87 rounds to 90, and the number 638 rounds to 600. The estimate is $90 \times 600 = 54{,}000$.

H

height   The straight-line measurement from the base of an object to the top. The measurements of a rectangular container include the length, the width, and the height.

K

kilogram   The standard unit of weight in the metric system. A kilogram is a little more than 2 pounds.

L

label   A word or abbreviation used to identify the unit of measurement of some quantity. A crate has a weight of 14 pounds. The label is *pounds*.

length   A straight-line measurement. The longer side of a rectangle is usually called the length.

liter   The standard unit of liquid measure in the metric system. A liter is about the same as 1 quart.

M

mean   Another word for *average*. A sum divided by the number of items that make up the sum.

measurement   A dimension, quantity, or capacity. The measurements of a room usually include the length, the width, and the height.

meter   The standard unit of length in the metric system. A meter is a little more than 1 yard.

metric system   A standard of measure based on tens, hundreds, and thousands. The standard unit of length in the metric system is the meter. The standard unit of weight is the kilogram. The standard unit of liquid measure is the liter.

minuend   The number in a subtraction problem from which another number is subtracted. In the problem $9 - 2 = 7$, the minuend is 9.

minus   Reduced by some amount. For example, seven minus one is six. The minus sign ($-$) is the symbol used for subtraction: $7 - 1 = 6$.

multiplicand   The number in a multiplication problem that is multiplied by another number. In the problem $25 \times 4 = 100$, the multiplicand is 25.

multiplication   A mathematical operation with whole numbers that consists of adding a number (the multiplicand) a certain number of times. For example, the problem $6 \times 3 = 18$ means finding the sum of three sixes: $6 + 6 + 6 = 18$.

multiplier   The number in a multiplication problem by which another number is multiplied. In the problem $12 \times 4 = 48$, the multiplier is 4.

O

operation   A mathematical process such as addition, subtraction, multiplication, or division.

P

parallel   Being an equal distance apart. The sides opposite each other in a rectangle are parallel.

partial product   In a multiplication problem, the result of multiplying the multiplicand by one digit of the multiplier. In the problem 125×13, the partial product of multiplying 125 by 3 is 375.

perimeter   A measure of the distance around a flat figure. A rectangle that is 3 feet long and 2 feet wide has a perimeter 10 feet. $3 + 3 + 2 + 2 = 10$

place value   The number that every digit stands for. For example, in 235, the digit 3 stands for 30 because 3 is in the tens place. The digit 2 stands for 200 because 2 is in the hundreds place.

plus   Increased by some amount. For example, two plus seven is nine. The plus sign ($+$) is the symbol used for addition: $2 + 7 = 9$.

product   The answer to a multiplication problem. In the problem $3 \cdot 6 = 18$, the product is 18.

Q

quotient   The answer to a division problem. In the problem 63 ÷ 9 = 7, the quotient is 7.

R

rate   An amount whose unit of measure contains a word such as *per* or *for each*. For example, the speed (or rate of speed) of a moving vehicle is often measured in miles per hour.

rectangle   A four-sided flat shape with four square corners. A page of a newspaper is a rectangle.

remainder   A number left over when one number divides into another. The answer to 25 ÷ 6 is 4 with a remainder of 1.

rounding   Making an estimate, ending in zeros, that is close to an original value. 83 rounded to the nearest ten is 80. 1,924 rounded to the nearest thousand is 2,000.

S

short division   Mentally multiplying and subtracting in a division problem. This division method works best with one-digit divisors. For example, 1,302 ÷ 6 becomes:

$$6\overline{)1,3\overset{1}{0}\overset{4}{2}}\ \ ^{217}$$

square   A four-sided flat figure with four right angles and four equal sides.

square units   A category of measurement for area. The area of a floor is measured in square feet or square yards or square meters.

subtraction   The mathematical operation used to find the difference between two numbers. The problem 14 − 5 = 9 is an example.

subtrahend   The number in a subtraction problem that is subtracted from another. In the problem 8 − 5 = 3, the subtrahend is 5.

sum   The answer to an addition problem. In the problem 4 + 9 = 13, the sum is 13.

symbol   A printed sign that represents an operation or a quantity or a relation. The symbol + means "add." The symbol ° means "degree." The symbol = means "is equal to."

T

table   An orderly arrangement of numbers in rows and columns. Train schedules are often in the form of a table.

time   A measurement from a point in the past to a more recent point. The units of measurement for time are seconds, minutes, hours, days, weeks, months, years, and so on.

total   Another word for *sum;* the answer to an addition problem.

U

units   The name of the right-most place in the whole number system. In the number 623, the digit 3 is in the units place.

units of measurement   Labels for determining a quantity. For example, pounds and ounces are units of measurement for weight. Meters and yards are units of measurement for length.

V

volume   The amount of space occupied by a 3-dimensional object. A rectangular box that is 4 feet long, 3 feet wide, and 2 feet high has a volume of 24 cubic feet.

W

whole number   A quantity that can be divided evenly by 1. For example, 3 and 14 and 2,500 are whole numbers. 2/3 and 4/5 are not whole numbers; they are fractions.

width   Usually the measurement of the shorter side of a rectangle. A 3-inch by 5-inch photograph has a width of 3 inches.

Z

zero   The mathematical symbol 0. When zero is added to any other number, the sum is that other number. For example, 5 + 0 = 5.

Index

A

Addend 15
Addition 15+
Area 137
Average 111, 149

B

Borrowing 39

C

Calculator, 26, 51, 77, 109, 162
Carrying 19, 71
Commutative property
    of addition 99
Commutative property of
    multiplication 99
Compatible pairs 24, 163
Cubic units 140

D

Difference 35, 52
Digit 6
Dividend 90
Division 90+
Divisor 90

E

Estimating 25, 50, 76, 100, 107,
    109, 178

F

Factor 62
Front-end rounding 76, 165

H

Height 140

L

Length 135, 137, 140

M

Mean 111, 149
Mental math 15, 35, 62, 90, 99, 163
Metric system 168
Minuend 35
Multiplicand 62
Multiplication 62+
Multiplication table 65
Multiplier 62

P

Partial product 67, 69
Perimeter 135
Place value 6, 9
Product 62

Q

Quotient 90

R

Rectangle 135, 137
Regrouping 19, 39, 71
Remainder 96
Rounding 11, 25, 50, 76, 107, 165

S

Short division 94
Square units 137
Subtraction 35+
Subtrahend 35, 49
Sum 15, 27

T

Total 15, 27
Two-digit accuracy 109, 167

U

Units 6
Units of measurement 130–134, 168
U.S. customary system 168

V

Volume 140

W

Width 135, 137, 140

Z

Zero 9, 45, 49, 74, 76, 99, 107, 163